H. Komiyama, S. Kraines

T0192272

Vision 2050
Roadmap for a Sustainable Earth

H. Komiyama, S. Kraines

Vision 2050

Roadmap for a Sustainable Earth

 Springer

Hiroshi Komiyama, Ph.D.
President
The University of Tokyo
7-3-1 Hongo, Bunkyo-ku
Tokyo 113-0033, Japan

Steven Kraines, Ph.D.
Associate Professor
Division of Project Coordination
The University of Tokyo
5-1-5 Kashiwa-no-ha, Kashiwa
Chiba 277-8568, Japan

ISBN 978-4-431-99803-7
eISBN 978-4-431-09431-9

Springer is a part of Springer Science+Business Media
springer.com
Printed in Japan

Printed on acid-free paper

Preface

Can we humans continue to live and work as we have until now within the resource limits of the earth? And can we sustain the earth's bountiful resources, including a clean and healthy environment, for generations to come? Recently, alarms have been sounded predicting a catastrophic future for the earth's environment and resources, and most informed people feel anxious about the dangers that may lie ahead. However, few of the people sounding these alarms have offered convincing plans of how we can navigate safely past the impending dangers. The goal of this book is to propose a concrete vision of a road to a sustainable future for humanity and the earth. By a "sustainable earth," we mean a way of living our lives and conducting the various activities that support our lifestyles within the bounds of the earth in such a way that we do not exceed those bounds, either by depleting non-renewable resources or by overloading the capacity of the earth and particularly the earth's biosphere for renewal. As we will see in this book, the sustainability of the earth is a dynamic process of circulations in large-scale and complex systems. Human society is one such system, and in order to make human existence on the earth sustainable, we must figure out how we can create a social infrastructure that sustains circulations matching those of the earth.

This book will show how – by virtue of science and technology – we can create an infrastructure for conserving energy and recycling materials by the year 2050. Furthermore, this book will show how that infrastructure will put us on the path towards maintaining high standards of living without depleting the earth's resources or despoiling the environment. Realizing this infrastructure will require that we establish a good relationship between society and technology. This relationship must be based on clear and honest communication between researchers in technology and stakeholders in society.

Since ancient times, human beings have developed and improved technologies: making tools, mastering fire, learning to plow the land. In tool-making, humans have progressed from shaping implements from stone, pottery, bronze and iron to manufacturing synthetic fibers and high-tech ceramics. In harnessing sources of energy, we have gone from burning wood to releasing the power of coal, oil, natural gas and nuclear energy. To improve agricultural yields, we have progressed from letting fields lie fallow to spreading manure to synthesizing chemical fertilizers. As a result of these technologies, human beings have flourished and populations have swelled. Although poverty remains a serious global problem, most people today, even in the developing world, live lives of health, wealth, comfort, and convenience unimaginable to our ancestors.

But impending depletion of resources and degradation of the environment have begun to threaten the civilization we have achieved. The seemingly boundless sky and vast ocean – which once seemed capable of absorbing every waste we threw out or spewed out – are now changing dramatically as a result of human activity. It is now obvious that the earth is but one small planet of limited size and resources. There are already clear indications of the serious problems posed by depletion of energy resources, by global warming, and by the massive generation of waste products. If we do not make changes in the way we use and reuse the earth's resources by the middle of the 21st century, these problems threaten to swamp the ship of human civilization.

As the negative side effects of our material civilization have become increasingly obvious, many people have begun to question our modern lifestyle. Awakened to the immensity of the garbage problem, the global warming problem, or some other threat to human civilization, many have come to feel that they must take action. If separating the garbage will help, many are prepared to do so. If solar energy is the solution, many who could afford it would be willing to install photovoltaic solar cells on their roofs. But one reason that people fail to follow through on these good intentions is that they are unsure what effect their efforts will actually have on global problems. In fact, many of us are doubtful whether our individual efforts will have any effect at all. As a result, many who fear for the environment and want to take action instead hesitate and end up doing nothing.

It is true that a variety of actions have been initiated that are intended to achieve a sustainable earth. Recycling is one example. Yet we still hear some experts claim that the cost of recycling makes it unrealistic or even that it is more harmful to the environment to recycle than not to recycle certain products. Some experts claim that solar cells are the energy trump card of the 21st century, but others say that such technologies are too expensive, and moreover they would scarcely contribute at all to the mitigation of the potential energy crisis. To take the first steps towards a sustainable earth, we need answers to these conflicting claims. More important, we need a comprehensive vision we can all share of what human civilization must look like at some point in the future for the sustainability of the earth to be assured. With such a shared vision, we could clearly evaluate the roles to be played by technologies such as solar cells and activities such as recycling.

The goal of this book is to lay out a comprehensive vision of how we could work together to put our society on the path toward sustaining a high quality of life on a planet with limited resources, and of the concrete steps we must take to get there. The 21st century is a crossroads where humanity will decide whether to take the path towards a sustainable society or the path towards environmental degradation and resource depletion. With this choice in mind, this book will submit "Vision 2050," a comprehensive vision aimed at reversing the trend toward resource depletion and environment degradation by 2050. "Vision 2050" is a concrete plan for a society based on recycling of materials, renewable energy, and energy efficiency that can be achieved by the middle of the 21st century and that would put us on a path to a sustainable earth by the 22nd century. By making "Vision 2050" a reality, we should be able to safely navigate past the trilemma of depletion of oil resources, global warming, and massive generation of wastes, to achieve a social foundation for supporting the sustainable development of humanity.

To make the earth a sustainable foundation for human life, we must reduce the burden that we place on it. Re-evaluating our modern material lifestyle is certainly important. But will it be enough? Today's global human population of 6.6 billion is predicted to reach 9 billion by the middle of the 21st century, and inevitably material consumption in the developing world will increase dramatically as a result. Because this population explosion will place a huge and ever-increasing burden on the earth's resources, it is clear that just changing lifestyles will not be enough to achieve a sustainable earth. We must consider how we can further reduce the burden of humanity on the earth. One way to do this is by developing technologies to reduce the inflow of natural resources and the outflow of waste materials accompanying each unit of human activity. And as this book will demonstrate, the impact of such technologies can be tremendous.

Vision 2050 is a concrete proposal for how we can resolve the problems of an imperiled environment and shrinking resources while still enabling all peoples on the earth to achieve living standards enjoyed by those in developed countries today. Vision 2050 is based on three necessary conditions: 1) increasing the efficiency of energy use, 2) increasing the recycling of materials in manufactured goods and infrastructure (what we will call "human artifacts") and 3) developing renewable sources of energy. Through the realization of an efficient recycling society, these conditions should be attainable. The key to achieving this kind of social infrastructure is establishing a circulation system from waste products to raw materials that takes over some of the burden that we are currently putting on the earth's biosphere.

This book will show that the goal of creating an energy-efficient, recycling society is possible in part because our legacy from the 20th century is not all negative. Certainly the 20th century has left us many problems to clean up, such as pollution of the land, air and seas. Nearly all of the infrastructure and manufactured goods around us – buildings, railroads, highways, cars and household appliances – must be disposed of in the 21st century, a casting off that could result in a huge burden on the earth. However, under certain conditions, it is possible for us to consider these human artifacts as a positive inheritance even after they have reached the end of their intended use. In most of the world, human artifacts – that is all of the things that we manufacture – will approach a state of "artifact saturation" by the middle of the 21st century. This book will show that we can use technology not only to develop large-scale sources of renewable energy and to revolutionize our energy efficiency, but also to recycle almost all of the materials in the waste products from the previous century, thereby reducing the use of natural resources for manufacture of new products to near zero.

It cannot be denied that the twin titans of science and technology have given human beings the potential to destroy ourselves. But if we develop science and technology wisely, we can use them to create a sustainable environment supporting a comfortable lifestyle in a clean and beautiful planet that humanity can enjoy for generations to come. Therefore, we need to make the correct choices concerning the direction of technology, and these choices can be made and implemented only through the consensus of society. There has never been a time when a good relationship between society and technology has been more important.

The rest of the book is laid out as follows.

Chapter 1 explains the mechanisms by which the circulation system of the earth's biosphere has been sustained by the energy of the sun until now. In this chapter, we will examine the way in which human activities have been disrupting this circulation by considering the global life cycle of the basic materials used to produce human artifacts. Throughout, we will clarify the nature of the three potential world-wide catastrophes of "global warming," "fossil fuel depletion," and "massive generation of waste" – catastrophes that will occur if we continue to act as we have.

In **Chapter 2**, we will see the ways in which we consume energy for the two basic activities of "making things" and "daily life." We will need to study some of the subtle concepts of energy, particularly the law of conservation of energy, in order to explain why, despite the physical law that energy cannot be destroyed, the potential crisis of "depletion of energy" is real. Chapter 2 attempts to do this using non-scientific language and examples from everyday life. Finally, we will see how we can extend the lifetime of our current energy resources by increasing energy efficiency.

In **Chapter 3**, for each of the activities that contribute significantly to the consumption of energy by humanity, including manufacturing processes in "making things" and human activities in "daily life," we will see what the minimum amount of energy is that must be consumed in the ideal case. From these ideal energy consumption rates, we will estimate the minimum energy required for all of the people in the world to attain a living standard equal to that currently enjoyed by those in developed countries. This will give us a theoretical target for the reduction of energy use that can be attained through technology.

Chapter 4 compares the limits for energy consumption rates estimated in Chapter 3 with what is attainable by the current state of technology for human activities in "daily life." Specifically, we will examine the potential to improve the efficiency of automobiles as well as of energy-consuming appliances in homes and office buildings, such as air conditioners. Finally, we will take a look at the state of the art in technology for generating electric power in conventional thermal power plants and discuss what we can expect in the future.

Chapter 5 begins to lay out a path towards creation of a social infrastructure based on the recirculation of basic manufacturing materials by recycling. In particular, this chapter will demonstrate, both in theory and through analysis of the current situation in society, that using recycled materials for manufacturing is not only technologically possible but also economically sound because it will significantly reduce energy consumption.

Chapter 6 considers the types of energy resources that are potentially available for replacing non-renewable fossil fuels. This chapter will show us the current state worldwide in the use of renewable energy sources, such as solar cells, wind turbines, and geothermal energy generators, and it will outline possible future scenarios for implementing large-scale systems for generating energy, systems based on the most promising of the renewable energy sources.

Drawing together the discussions from the previous chapters, **Chapter 7** puts forth "Vision 2050" as a comprehensive roadmap for global sustainability that could realistically be achieved by 2050.

Chapter 8 looks at the synergistic relationship between society and technology that is needed to make the right decisions among the various choices for the future within the framework of Vision 2050. Several new approaches based on emerging technologies for helping to realize this synergy are introduced, focusing particularly on structuring expert scientific knowledge and sharing that knowledge in ways that are most beneficial and accessible to the people who can apply it towards the achievement of a sustainable human existence on the earth.

Contents

Chapter 1
Is the Earth Sustainable?

1 Changes from Which the Earth Recovers, and Changes from Which the Earth Does Not

The Continuous Renewal of the Circulating Earth

> *"Flowers bloom alike, year after year. But not people."*
> *(Translation of an ancient Japanese proverb)*

For millennia, human beings never questioned nature's continuous renewal. Each year the seasons changed, but as spring rolled round again, the same trees blossomed and bore fruit. Until today, humans have lived their lives assuming that this circulation of nature would always continue.

In spring, plants use the energy of sunlight to absorb carbon dioxide (CO_2) from the atmosphere together with water from their surroundings to produce roots, stems, branches, and leaves. This process is called photosynthesis. Through spring and summer, as land plants flourish around the world the amount of CO_2 in the atmosphere decreases. When those plants lose their leaves in the autumn, the fallen leaves are eaten by insects and other animals. A part of this is oxidized into CO_2 when those animals respire; that is, the leaves are breathed out as CO_2. The leaves that are not eaten, together with the feces and dead bodies of the animals, become organic matter in the soil. That organic matter is used by microorganisms and other denizens of the soil and eventually transformed back into CO_2. So after several years, all of the CO_2 from the atmosphere that was taken up by a plant during its lifetime is returned to the atmosphere. Carbon circulates around the earth in this way, and each year the earth has returned to its original state.

Like all other living things, humans have lived out their lives within the circulations of the earth. Agriculture is one human activity that traditionally has been relatively well adapted to the cycles of nature. If rice seedlings are planted in the

Hiroshi Komiyama and Steven Kraines
Vision 2050: Roadmap for a Sustainable Earth.
© Springer 2008

rice fields in the spring, rice can be harvested in the autumn. After the rice plants are cut down and the rice is harvested, winter comes and the fields become desolate. However, if rice is planted the next spring, an abundant harvest will come again the following autumn. Fishing is another such activity. Even if pre-industrial fishermen took in large catches of salmon from early summer into the autumn, at the beginning of the next summer, the salmon would return.

The earth has always been a place of dynamic changes. But because it has always returned to its original state after each year, the earth has provided a reliable stage for human civilization.

Recently, though, this pattern of continuous renewal has started to derail. Our planet is being affected by continuous and dramatic changes – changes from which it does not recover each year.

Changes from Which the Earth Does Not Recover

One change from which the earth does not recover is the rising level of CO_2 in the atmosphere (see figure 1-1). For at least the last thousand years, the yearly average concentration of CO_2 in the earth's atmosphere remained nearly constant at 280 ppm (in volumetric terms). However, in the 19th century, that concentration began to rise, and during the second half of the 20th century, the rate of increase has accelerated dramatically. The concentration of CO_2 in the atmosphere at the end of 2007 was about 384 ppm. And if the CO_2 concentration continues to increase at the current rate, it will be double the pre-industrial concentration of 280 ppm by the end of the 21st century. Actually, because the rate of increase itself is increasing, this doubling of the CO_2 concentration may occur even earlier.

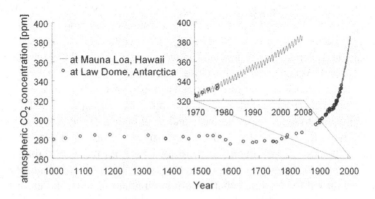

Fig. 1-1: Atmospheric CO_2 concentration from 1000 to 2008 (Data from National Oceanic and Atmospheric Administration: Dr. Pieter Tans, NOAA/ESRL and D.M. Etheridge et al., 2001, Law Dome Atmospheric CO_2 Data, 1GBP PAGES/World Data Center for Paleoclimatology Data Contribution Series #2001-083. NOAA/NGDC Paleoclimatology Program, Boulder CO, U.S.)

The increase in the concentration of CO_2 is not likely to be directly harmful to humans and other living things. In fact, there is some evidence that plant growth is being enhanced by the increase and that as a result forests are becoming greener and more lush. However, the increased concentration of CO_2 in the atmosphere is thought to be indirectly changing the circulations of the earth – changes that could have far more serious impacts on human civilization than the increase in plant growth. Specifically, the increase in CO_2 concentration is believed to be inducing global warming.

We know for a fact that the average surface temperature of the earth is increasing. However, because the earth's temperature varies greatly with location and time of year, it is difficult to measure the average temperature of the earth reliably. Furthermore, the temperature of the earth is affected by sun spots and other solar activity. Even the eruption of a large volcano can affect the earth's temperature because the dust that is exploded into the atmosphere during an eruption reflects incoming sunlight, reducing the amount of sunlight that reaches the earth's surface. Many factors such as these affect our measurements of the earth's temperature and make it difficult to determine the relationship between CO_2 and temperature. However, techniques for assessing this relationship have become more and more accurate. According to the latest investigations by scientists at the IPCC (Intergovernmental Panel on Climate Change) reported in 2007, a rise in the average surface temperature of the earth of 0.74°C has occurred already. The major cause of this temperature rise is believed to be global warming from the increase of CO_2 in the atmosphere that has occurred over the past century.

How Long Does It Take for Ice to Melt?

One result of global warming that is raising fears is the rise of the sea level. According to the 2007 IPCC report, the current rate of sea level rise is 3.1 mm per year. At this rate, the sea level will rise nearly 12 cm by 2050. More alarming is the possibility that large parts of the ice currently land-locked in Antarctica and Greenland will slide into the ocean. Although ice is less dense than sea water, if large land-moored ice shelves break off into the ocean, they will raise the sea level. The ice will displace the water around it the same way that putting ice cubes in a full glass will cause it to overflow. Experts estimate that if all of the ice in Greenland were to slide into the ocean, the sea level would rise more than 600 cm. On the other hand, in the same way that a full glass of ice water will not overflow even if all of the ice in the glass melts, the ice in the Arctic, which is already in the water, will not increase the sea level much, even if it melts.

The fact that global warming will cause a rise in sea level is relatively well-known. And you might think that if we stabilized the CO_2 concentration in the atmosphere, the sea level would stop rising. But this is not true. The rise in sea level results from the melting of land ice in places like Antarctica and Greenland as well as from the thermal expansion of sea water as the temperature of the oceans increases. And it takes a long time to melt large chunks of inland ice and raise the temperature of entire oceans.

Little pieces of ice, such as shaved ice, melt quickly, and a piece of ice the size of an icicle may take at most a day to melt. A chunk of ice the size of a glacier would take a much longer time to melt. If we assume that a glacier melts only from the outside, then with a melting rate of 1 cm per day, it would take 300 years for a glacier 100 meters thick to melt. Heating an entire ocean also takes centuries. Even if we can stabilize the surface temperature of the earth at some level above its the pre-industrial temperature, glaciers will continue to melt bit by bit, and the temperature of the oceans will continue to increase little by little. As a result, the sea level will continue to rise until the oceans can absorb the excess CO_2, the atmospheric CO_2 concentration can decrease, and the earth's temperature can begin to return to its current value. This may take centuries.

Global warming caused by the increase in the concentration of CO_2 in the atmosphere and the resulting rise in sea level are only two examples of how the earth is beginning to change in ways from which it cannot recover through its annual cycles.

So why is the earth unable to recover in the way that it used to? To answer this question, let's look into the framework by which the earth has repeated its cycles of yearly recovery until now.

2 Mechanisms for Recovery

Circulating Ecosystems Powered by the Sun

In 1998, there was a huge forest fire in Indonesia. This fire burned for several months, and satellite images showed that smoke from the fire extended as far as the Malay Peninsula. The smoke from this vast fire is even believed to have caused an airplane crash killing all 234 people on board. Although a fire of this size is rare, forest fires occur each year around the world. However, once a fire is extinguished, even the fire in Indonesia, plants grow back and the forest recovers. After a forest fire, plant life in the form of seeds and underground shoots remain in the soil, and when spring comes around again, the greenery returns to the forest. A forest fire can even be a good thing for a forest ecosystem as it rids the forest of dead wood and parasites. In fact, one reason given for the ancient custom of burning the dead leaves on the *Wakakusa* Mountain in Nara prefecture of Japan every January is that it helps to preserve the plant life on the mountain. Therefore, even forest fires are a part of the circulations of the earth's biosphere.

Another example of nature's recovery can be seen in the fishing industry. If not fished into extinction, salmon, tuna, mackerel and other species of wild fish will restock a fishery year after year because uncaught the adult fish spawn and produce juveniles that grow in turn into adult fish. But this growth requires food. And the food chain in the ocean begins with phytoplankton. Like land plants, phytoplankton grow through photosynthesis. Many of them are captured by zooplankton, which are eaten by little fish, which are eaten in turn by bigger fish. When we get to the source of the food chain in the ocean, we find that it is photosynthesis using energy from the sun. A similar food chain occurs on land. Through photosynthesis, land plants grow foliage and bear fruit, which herbivores eat to grow and multiply. Carnivores prey on the herbivores to sustain themselves, and at the same time they keep the numbers of herbivores in check.

In summary, the basis for the cycles of life in the ecosystems on land and in the sea is photosynthesis, a process powered by the energy of the sun.

The Wind and Rain Also Are Caused by the Sun

In addition to these ecosystem cycles that are sustained by photosynthesis, weather-related phenomena such as wind and rain are also powered by the sun's energy. Rain happens when water on the land and the sea is heated by the sun, evaporates, forms clouds, and coalesces into droplets that fall as rain. After the rain falls to the earth, it soaks into the ground and feeds little creeks that feed into larger streams. Ultimately, these merge into rivers that flow into the oceans. In this way, water circulates on the surface of the earth, driven by the energy of the sun.

Wind is created when air flows from high pressure zones towards low pressure zones. Low pressure zones are regions where the sun has heated the air making it rise, and high pressure zones are regions that are relatively less heated. In fact, the energy of the sun is the source of all the forms of air circulation, including trade winds, typhoons, seasonal winds, and even local breezes.

Both rain and wind play important roles in the biosphere. As water circulates by falling as rain, gathering into rivers, and flowing into the oceans, it dissolves nutrients from rocks and soil. Those nutrients are absorbed by plants during photosynthesis, taken up by animals when they eat the plants, and returned to the ground and water when the animals urinate or pass feces. Winds transport a variety of materials, including seeds and nutrient-laden dust. Together with photosynthesis by plants, these are the phenomena upon which the circulations of ecosystems are based, and they all are powered by the energy of the sun.

The Amount of Elements in the Biosphere Is Constant

The part of the earth where all of these ecosystem cycles occur is called the "biosphere." The biosphere is completely contained within a thin shell about 20 km thick, from the peak of Mount Everest to the bottom of the Mariana Trench. To get a feel for how thin the biosphere is, try drawing a circle on a letter size piece of paper to represent the earth. No matter how sharp you make your pencil, the line that you draw will be thicker than the biosphere. Almost all human activity occurs within this single thin layer.

It may surprise you to learn that for over ten million years, the total amount of each chemical element in the biosphere has hardly changed at all. Chemical elements, such as carbon, oxygen and hydrogen, are neither created nor destroyed during the normal processes that occur on the earth's surface. For example, CO_2 is changed into carbohydrates by photosynthesis; however, the amount of carbon in the carbohydrates is the same as the amount that was in the CO_2. That is what scientists mean when they say that chemical elements are conserved during chemical reactions.

The only case in which chemical elements are not conserved is when the atomic nucleus is changed in a nuclear reaction. In a nuclear reactor, the nucleus of a chemical element called uranium is changed and a different element such as plutonium is created. Even in nature, forces such as cosmic rays can cause one chemical element to change into another chemical element. However, this amount is insignificant. Conservation of mass, and of chemical elements in particular, is one of the fundamental principles upon which science is based. (Another is conservation of energy, which will be introduced in Chapter 2.)

Although the chemical elements are conserved in constant amounts, we have seen that they are changed into various forms as they circulate through the biosphere driven by the energy of the sun. For example, nitrogen in the atmosphere, which occurs as a molecule containing two atoms of nitrogen, N_2, is taken up by nitrogen fixing bacteria living in the roots of plants and transformed into ammonia. Some of the ammonia is taken up by the plant, which converts it into proteins. The plant protein is consumed by animals, and some of the nitrogen consumed is excreted by the animals in the form of urea. Bacteria in the soil consume the urea and produce an oxidized form of nitrogen called nitrate. Other bacteria consume the nitrate and convert it back into N_2, thus completing the cycle. All of the other chemical elements in the biosphere follow the same kinds of circulations, eventually returning to their original state.

But changes from which the earth does not recover, changes we saw earlier in this chapter, are beginning to occur in this very same biosphere. Why has this happened? What has suddenly interrupted the cycles of the biosphere, cycles that have returned the earth to its original state each year for thousands of years? In the next section, we will take a look at what has changed in the last century.

3 A Massive Intervention by Humanity into the Biosphere

A Century of Expanding Human Activities

In this section, we will look at three graphs illustrating how much human activities expanded in the 20th century. The first graph shows the total human population on

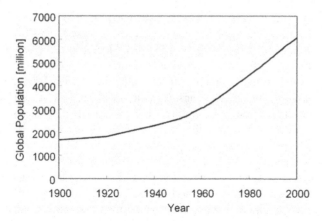

Fig. 1-2: Global population from 1900 to 2000 (Data from UN Common Database, United Nations Statistics Division)

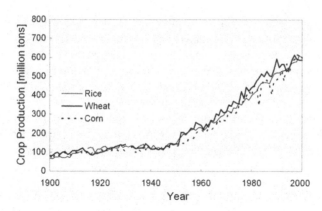

Fig. 1-3: Global production of the three major grains from 1900 to 2000 (Data from FAOSTAT database, Food and Agriculture Organization of the United Nations, the UN Common Database, United Nations Statistics Division, and B.R. Mitchell, International Historical Statistics, Palgrave Macmillan)

earth from 1900 to 2000 (figure 1-2). The human race entered the 20th century with 1.6 billion members and grew to 6 billion by the end of the century, an increase of almost four-fold. We use "billion" in the American English sense of one thousand million or 1,000,000,000.

The second graph illustrates how the production of agriculture has grown during the same period of time (figure 1-3). The production of agriculture as represented by the three major cereal grains – rice, wheat and corn – increased seven-fold. Because human population increased only four-fold, the average consumption of grain per person nearly doubled. The expansion of farmland area was one factor in creating this dramatic increase. However, particularly since the 1960's, the increase in agricultural production has been mainly due to increased yield from the same sized area. For example, from 1960 to 1995, the agricultural yield increased by 2.5 times. The main reason for this increase in efficiency is that a technique for manufacturing nitrogen fertilizer, which until the 1960's had to be obtained through nitrogen-fixing plants such as soybeans and other legumes, was successfully developed by synthesizing ammonia from nitrogen. However, this increase in efficiency may come at a cost. Experts say that in many parts of the world the large-scale agriculture made possible by the introduction of artificial fertilizer has seriously degraded the soil and therefore the ability of the land to produce the same agricultural yields each year. We may be getting some of our increased land productivity today at the cost of productivity in the future.

Fishery yields also increased as small fishing boats, which had been restricted to trawling the shorelines, were replaced with large ships that could fish the open seas. Furthermore, fishing nets and other equipment were improved, allowing the scale of fishing operations to become even bigger. However, these improvements in fishing practices meant that the fisheries were no longer able to completely recover each year. For example, when whaling was restricted to the shorelines, the cycles of nature could sustain the numbers of whales. But when whaling ships moved out into the Antarctic Ocean and began to hunt whales on a large scale, the numbers of whales diminished so much that concerns were raised that some whale species might become extinct. According to the State of World Fisheries and Aquaculture 2006 report of the UN Food and Agricultural Organization, over three quarters of the world fish stocks are being over fished.

The third graph shows production levels of iron and aluminum, two representatives of basic materials used to make the various goods and infrastructure components (figure 1-4). In the 20th century, production of steel increased twenty-fold,

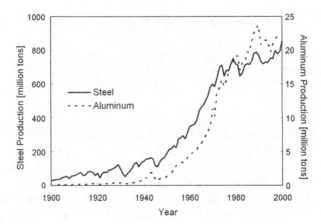

Fig. 1-4: Global production of iron and aluminum from 1900 to 2000 (Data from UN Common Database, United Nations Statistics Division and B.R. Mitchell, International Historical Statistics, Palgrave Macmillan)

and production of aluminum increased four thousand-fold. In fact, the production levels of almost all basic materials have increased from more than ten fold to several thousand-fold during the last century. Materials such as plastics and synthetic fibers did not even exist in the 19th century. Thus, the expansion of manufacturing and manufacturing-related human activities in the 20th century was particularly remarkable. And as we will see later in this chapter, the pressures of mining for resources and providing energy for manufacturing have also begun to disrupt the natural circulations in the biosphere.

There is a well-known equation among experts studying the sustainability of human existence on the earth. The equation states that the impact of humans on the earth equals the product of the human population, the affluence of that population as measured by the products and services consumed per person, and the impact on the earth of providing one unit of product or service. For example, the impact of food consumption is the human population times the average amount of food consumed per person times the amount of natural resources, such as water and land, needed to produce a given amount of food. The last factor in the equation – the size of the impact of providing a product which reflects the state of technology – is the inverse of the efficiency of the process providing that product. Since efficiency determines the factor in the equation where technology can play a role, it will be a major topic in this book.

Over the last few centuries, as the world's population has grown and the average per-person consumption of food and manufactured products has increased, the human impact on the biosphere has increased by orders of magnitude. Just in the last decade, human population has increased 10%, CO_2 emissions have increased about 25%, and production of basic materials such as iron and cement has nearly doubled. As a result of this impact, the biosphere is no longer able to return to its original state each year. In the next few sections, we shall look at human activities and the burdens each kind of activity imposes on the biosphere.

The Use of Fossil Fuel Resources

Human activities require energy. Once, this energy was obtained mainly by burning wood. However, as human activities expanded, wood burning was no longer enough to meet our energy needs. For example, charcoal was originally used in making iron. At that time, England was the leading producer of iron. But as a result of reckless lumbering to produce charcoal, the forests in England were so rapidly depleted that in the 16th century, Queen Elisabeth I had to issue restrictions on the logging of forests. Thereafter, the iron industry in England declined, and countries richer in forests, such as Russia and Sweden, were able to become iron exporters. The reason that England could reclaim her hegemony in iron production during the industrial revolution was because of coal.

The use of coal resulted in an expansion of industry. But later coal was eclipsed by oil as the star of the energy show. Oil has higher energy content per ton than coal. Furthermore, because oil is a liquid, it is easier than coal to handle during extraction, to load onto ships, and to fill into combustion furnaces. The explosive expansion of industry in the latter half of the 20th century was made possible by the large-scale use of oil. However, the use of the fossil fuels coal and oil, and later natural gas, has come at the cost of unprecedented impacts on the biosphere. The reason is as follows.

Fossil fuels are composed mainly of carbon and hydrogen. When fossil fuels are burned with oxygen from the air, CO_2 and water are released as by-products. However, the CO_2 and water produced by burning fossil fuels contain carbon and hydrogen atoms that had been buried deep underground and therefore had not been involved in the circulation of chemical elements in the biosphere. In other words, the CO_2 and water released by burning fossil fuels is matter added by humans to the constant amount of elements being circulated through the earth's ecosystem by the energy of the sun. Furthermore, this new matter is added to the atmosphere, a medium which circulates more rapidly than the other parts of the biosphere, such as the ocean. The amount of water added through the burning of fossil fuels is insignificant in comparison to the total amount of water in the earth's atmosphere, but the increased amount of CO_2 can no longer be ignored.

According to the 2007 report of the IPCC, the increase in the concentration of CO_2 that was shown in figure 1-1 is caused by the enormous production CO_2 through the burning of fossil fuels together with a similarly large amount of CO_2 generated through the cutting down of forests. When forests are cut down, the felled trees will eventually be turned into CO_2. The amount of CO_2 produced each year by burning fossil fuels is estimated to be 7.5 billion tons in carbon units, which we will abbreviate as "tons-C." It is important to make this distinction, because the mass of a carbon atom is only about a quarter of the mass of CO_2. In this book, when we are talking about amounts of carbon-based materials such as CO_2 and fossil fuels, we will always use this measure of tons-C. The amount of CO_2 generated through the cutting down of forests is believed to be about 2.3 billion tons. When other emissions of CO_2 by human activities are added in, the total amount of CO_2 emitted each year through human activities is more than 10 billion tons.

The amount of CO_2 in the atmosphere at the end of the 20th century was about 700 billion tons, so human activities are increasing the CO_2 content in the atmosphere by more than 1% each year. A continuous annual increase in the atmospheric CO_2 concentration of this magnitude has never before been experienced in the history of human civilization. If we continue to emit CO_2 at the current rate, by the end of the 21st century, we will double the amount of CO_2 in the atmosphere today. Some portion of the 10 billion tons of CO_2 emitted into the atmosphere gets redistributed to the other parts of the biosphere. About half is absorbed by the oceans or taken up by new growth in the forests. The other half accumulates in the atmosphere. Therefore, the concentration of CO_2 in the atmosphere is linked to fundamental conditions on earth such as the surface temperature, which controls the rate of absorption by the oceans, and the rate of photosynthesis, which controls the uptake of CO_2 by plants.

Diminishment of Nature and Accumulation of Human Artifacts

When we turn our attention to the realm of living things, the increasing number of species that have become extinct is alarming. It is reported that over 100 species per day, mainly insects, are disappearing from the face of the earth. The IPCC report estimates that as many as 30% of all plant and animal species face the possibility of extinction if global warming continues unabated. Of course the diversity of species should be treasured in and of itself, but there is also concern that a reduction in species diversity could reduce the resilience of ecosystems to disaster and disease. And once a species becomes extinct, it is essentially gone forever.

The decimation of forests, particularly tropical rain forests, is also remarkable. According to the 2007 State of the World's Forests Report of the UN Food and Agriculture Organization, the rate of deforestation is decreasing; nevertheless, 130,000 km^2 of forests are cut down every year. One result of this rapid loss of forests is that deserts are encroaching at an unprecedented rate on populated areas around the world, such as the Sahel Strip at the southern fringe of the Sahara Desert. The 2007 State of the World's Forest Report estimates that 135 million people may be forced to leave their homes as a result of desertification. For example, it is reported that sub-Saharan Africa loses 1% of the productivity of its agricultural land each year to the expanding desert.

As our natural resources are diminishing, human artifacts such as buildings, roads, and cars are rapidly accumulating. The accumulation of human artifacts in the biosphere started to become conspicuous in the 20th century. For example, although Tokyo has been a place where people have gathered since ancient times, most of the buildings, roads and cars we see there today were not there at the beginning of the 20th century. We can see this accumulation in figure 1-4. The area under the lines showing the rate of production of iron and aluminum indicates the total amount of material produced by a certain time. It is clear that most of the basic materials used in human artifacts, such as iron and aluminum, were produced in the second half of the 20th century.

As cities accumulate human artifacts, they are simultaneously disgorging huge amounts of waste. Recently, disputes have arisen around the world over the disposal of garbage. The fact is that the natural environment around cities is unable to absorb the massive amounts of waste we produce.

The Influence of Toxic Materials

Toxic materials produced by human activities, of which small amounts can wreak havoc on organisms and ecosystems, are also interrupting the cycles of the biosphere. Toxic materials have a long history. During the industrial revolution, for example, toxic materials contributed to the polluted and unsanitary conditions of the air and water in London. The danger of toxic materials and their effects on ecosystems became the focus of public debate in 1962, when Rachel Carson published *Silent Spring*. Japan, too, has suffered many environmental pollution incidents, including the heavy metal pollution from the *Ashio* copper mines, the mercury pollution in *Minamata* bay, and the air pollution at *Yokkaichi*. All of these incidents were the result of industrial emission of toxic materials.

Acid rain is a form of toxic pollution that transforms forests and lakes into barren landscapes. Acid rain is mainly caused by the combustion of fossil fuels. When fossil fuels are burned, sulfur in the fuels and nitrogen from the air combine with oxygen to create sulfur oxides and nitrogen oxides. When these compounds are emitted into the atmosphere, they react with cloud water to become strong acids, such as sulfuric acid and nitric acid. When the cloud water turns into rain, the sulfuric and nitric acids make the rain water highly acidic. This acid rain produces a range of adverse effects on ecosystems, buildings, and human health.

The damage caused by acid rain cannot be confined by borders between countries, making acid rain an international issue. At the time of the industrial revolution, sulfur and nitrogen oxides generated by burning coal in England were carried by the wind across the North Sea and ended up forming acid rain that caused damage to forests and lakes in Scandinavia. Similarly, in North America emissions from fossil fuel combustion at U.S. steel-making plants around the great lakes have caused extensive damage in Canada. And recently, reports have begun to appear that acid rain originating in China is influencing Korea and Japan.

Another example of toxic materials is CFCs (chlorofluorocarbons), often known by the brand name Freon. CFCs, which do not burn or change chemical form easily, are good cleaning agents. And they can easily be converted from a liquid to a gas and vice versa. When they hit the market in the 1930's, they were hailed as one of the best chemical compounds ever developed. However, these same chemical compounds are now known to be a major cause of the depletion of the ozone layer. In the ozone layer (which is a part of the stratosphere in the upper atmosphere) CFCs react with ozone resulting in the destruction of ozone molecules. Ozone in the stratosphere acts as a filter to absorb ultraviolet radiation in sunlight, radiation that would otherwise damage genetic structures in living cells. Thus there is concern that depletion of ozone in the stratosphere by CFCs will give rise to increased rates of skin cancer and other genetic disorders.

Many other problems related to a range of toxic materials – from residual agricultural chemicals to dioxins to endocrine disruptors – are now raising concern and drawing scrutiny.

The examples above make it clear that human activities are beginning to disturb the natural cycles in the biosphere. As summarized in figure 1-5, human activities transform mineral resources into artifacts such as manufactured goods and urban infrastructure. Some of these artifacts accumulate within a society. However, many are discarded back into the biosphere. As a result, the biosphere is being flooded with human artifacts that have ceased to be of use, together with the CO_2 generated from fossil fuels used to produce these artifacts and various toxic by-products. All of this waste spewed or tossed into the biosphere disturbs the workings of the biosphere.

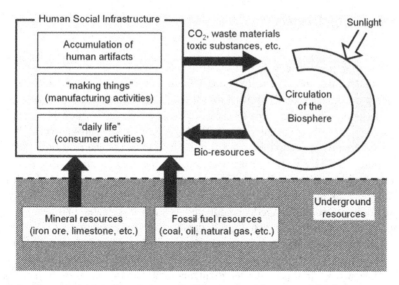

Fig. 1-5: The material interflows between the biosphere and human social infrastructure caused by human activities

4 The Flow of Materials Resulting from "Making Things"

Let's look at the picture shown in figure 1-5 from a different angle. Some human activities, such as agriculture and fisheries, make use of living resources indirectly derived from the sun. Therefore, as long as they are not carried out in excess, these activities do not cause damage to the circulation system of the biosphere. However, the activities of "making things" that involve the manufacture of artifacts using resources from underground are different. The reason is that activities of "making things" dig up materials that hitherto had been isolated underground and release them into the circulation system of the biosphere.

What materials are used to manufacture artifacts? Looking around, we see that paper and other wood products, metals such as iron and aluminum, non-metal minerals such as glass and concrete, and petroleum products such as plastics, rubber and synthetic fibers account for most of the materials used in human artifacts. The use of materials derived from animals, such as leather and shells, is miniscule in comparison.

In the following sections, we are going to look at the flow of these basic materials from when they are extracted from the earth as natural resources to when they are returned to the earth as waste. This is called the "lifecycle" of the materials, and it will give us a different perspective on the way human activities are disturbing the cycles of nature. In particular, we will see that there are three types of lifecycle: accumulation, one-way flow, and recirculation.

Accumulating Metals

First, let's look at the lifecycle of iron. Iron ore, the raw material for iron, is iron oxide – that is, iron bonded to oxygen. This iron ore is converted into iron in a huge reaction vessel, called a blast furnace or a shaft furnace, through the use of fossil fuels, mainly in the form of coke. Coke is a form of carbon produced by heating coal in the absence of oxygen. In the blast furnace, the carbon in coke bonds to the oxygen atoms, stripping them from the iron atoms in the iron ore, and producing pure iron. This chemical process is called the "reduction" of iron ore. The iron that is produced in a blast furnace is called "pig iron," and currently almost 900 million tons are produced each year worldwide. Pig iron is tempered with various additives, rolled, shaped, and cut; and its surface is treated in different ways to create the various iron and steel products that we see in the market.

A plant built around a blast furnace that carries out the entire process from reduction of iron ore to delivery of iron products is called an integrated iron and steel making works. In this integrated plant, about 600 kg of coke is used to produce one ton of steel. Because coke is made from coal, the production of each ton of steel consumes approximately 600 kg of fossil fuel resources.

As of the year 2007, the total production of iron worldwide has exceeded 1.3 billion tons. If only 900 million tons is from iron ore, where does the rest of the iron come from? When iron and steel products reach the end of their life spans, they are collected as scrap, melted down, and remanufactured to produce new iron and steel products. Globally, about 400 million tons of iron is produced from scrap. The fraction of total iron production that comes from scrap is therefore about one third. This fraction is often called the "recycle ratio," but this is a misleading expression. Saying that the recycle ratio is one third implies that two thirds of the iron is thrown away without recycling, but this is not the case. There is little accurate data on how much iron and steel is thrown away in garbage dumps, but it is thought to be far less than the amount recovered as scrap. Most of the difference between the amount of iron and steel that is supplied to the market and the amount that returns to the iron and steel making plants as scrap is accumulated in the infrastructure of society as artifacts.

Figure 1-6 shows a diagram summarizing the flow of iron in the biosphere. Many of the flows shown in the diagram cross international borders and oceans. Japan is a particularly good example because Japan has few natural resources and must import many of its raw materials from other countries. So let's look at a concrete example of the flow of iron ore from Brazil and coal from Australia to provide iron in Japan. Iron ore from Brazil is accumulated as iron in skyscrapers and highways in Japan, and coal that had been buried underground in Australia is released into the atmosphere as CO_2, where it contributes to the increase in global warming. Human artifacts eventually reach the end of their product lives, but most of the iron in them is made back into iron products. A small part of the iron is thrown away in garbage dumps, and over a long period of time, this iron rusts away and becomes

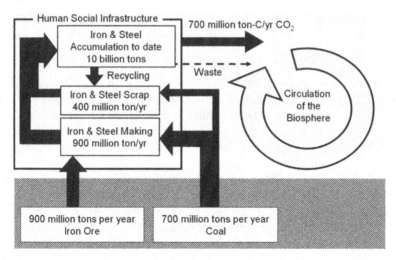

Fig. 1-6: The lifecycle of iron

iron oxide. This iron can be thought of as iron oxide that is transported from Brazil to a garbage dump in Japan. Similar flows occur between other producers and other consumers of natural resources for iron production. This is the lifecycle of iron in the biosphere, a lifecycle created by human activities.

Aluminum, the metal with the highest production level next to iron, is produced from ores comprised mainly of bauxite, or aluminum oxide. Because the bond between aluminum atoms and oxygen atoms is so strong, it is impossible to use carbon to remove the oxygen atoms through reduction as in the case of iron. Instead a different method is used. First, the bauxite is mixed with fluorides to reduce the melting point. Then the mixture of bauxite and fluorides is melted, and the molten bauxite is split into aluminum and oxygen through electrolysis.

The electricity used in this process accounts for nearly all the energy required to produce aluminum. And approximately 2% of the electricity generated in the world is consumed in producing aluminum. Countries like Japan, where the price of electricity is relatively high, do not produce their own aluminum. Instead, they import ingots of aluminum produced from bauxite in countries with cheap electricity, like the U.S. and Canada. Like steel, much of waste aluminum is recycled. The global production of aluminum from bauxite is more than 30 million tons per year, and the production from aluminum scrap is more than 10 million tons.

Let's take a look at the global flow of aluminum for use in Japan. Bauxite dug from mines in Australia is transformed into ingots of aluminum using hydropower in Indonesia, ingots which are then transported to Japan. This raw aluminum is made into products such as cans and window frames. And when those products are no longer needed, most of the aluminum contained in them is recycled into new products that are circulated back into the market. The portion of aluminum thrown

into garbage dumps is eventually converted back into aluminum oxide. So this
portion of the flow is equivalent to transporting bauxite from Australia to a garbage
dump in Japan.

The lifecycles of most metals currently operate in the same fashion as those
shown for iron and aluminum. Recycling of rare metals such as platinum, cadmium,
palladium, iridium, copper, and mercury has an even greater potential for making
society more sustainable. One reason is that rare metals tend to be more costly to
extract from natural resources. However, perhaps more importantly, rare metals are
often highly toxic, making it necessary to use expensive disposal methods if the
metals are not recycled.

The confirmed recoverable reserves of both iron and aluminum ore are large
enough that even if production is continued at today's levels, they would last for
two to three centuries. So we do not need to worry about depletion of these natural
resources for a long time. However, as you will discover in this book, if we continue
to use these natural resources to provide most of the basic materials that we use,
we will end up consuming tremendous amounts of energy and covering the earth's
surface in waste.

The One-Way Flow of Cement and Glass

Concrete and glass are the major non-metal minerals used in human activities. So
what do their lifecycles look like?

Concrete is sand and gravel bound together with cement. Cement is calcium
oxide formed when limestone is heated, driving off CO_2. About 100 kg of fossil
fuels are consumed in producing one ton of cement. Concrete is used to construct
buildings and highways, and most of the waste concrete generated when the buildings and highways are torn down is pulverized and used as low-grade materials in
applications such as roadbeds. However, the demand for these low-grade materials
is gradually decreasing. For example, in Japan of the total amount of 37 million tons
of concrete waste generated in 1995, more than 10% was not recycled. Almost all
of this concrete can be considered as having been thrown away in garbage dumps.

In short, the lifecycle of concrete unfolds as follows. Sand, gravel, and limestone
are collected from rivers and mountains, made into concrete through the use of
fossil fuels, and accumulated in the infrastructure of society. However, eventually
all this concrete becomes waste material. Here is what the lifecycle of Japanese
concrete looks like from a global perspective. Coal buried underground in America
and other parts of the world is transformed into CO_2 and released into the atmosphere. Sand, gravel, and limestone from the rivers and mountains of Japan are
accumulated in human artifacts such as buildings and highways. All of those artifacts are eventually torn down, and all of that concrete finally ends up in garbage
dumps. So we see that the lifecycle of concrete is essentially a one-way flow – from
the consumption of natural resources to burdens on the environment in the form of
expanding garbage dumps and increasing CO_2 in the atmosphere (figure 1-7).

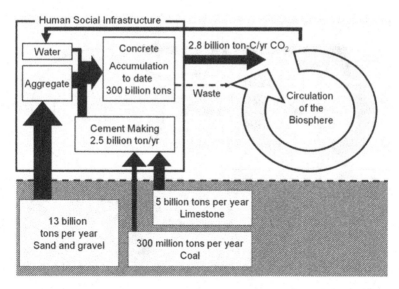

Fig. 1-7: The lifecycle of concrete

Glass products are formed by heating a mixture silicon oxide, sodium carbonate, and calcium carbonate to drive off CO_2, and then melting down, shaping, and solidifying the mixture. About 200 kg of fossil fuels are consumed in making one ton of glass. In Japan, the current recycle ratio of glass is about 50%, so on average glass from natural resources is used twice in manufactured products. However, in the end, the lifecycle of glass is almost the same as that of concrete. Silicon oxide, sodium carbonate, and calcium carbonate in quartz, soda ash, and limestone are collected from the rivers and mountains of Japan and other countries and eventually end up being transported to garbage dumps. At the same time, oil from places like the Middle East is emitted as CO_2 into the atmosphere.

Petroleum Products Are Also a One-Way Flow

Plastics and synthetic fibers are examples of large molecules called polymers that, unlike the molecules of CO_2 and nitrogen, are composed of long strings of atoms – strings ranging from tens to millions of atoms. Currently, the production of polymers worldwide is more than 200 million tons, and on average about two tons of oil is used to make one ton of plastic. Plastic is a special product in the sense that oil is used both as a raw material and as an energy source for manufacture. To produce one ton of plastic, almost equal amounts of oil are used as energy and as raw material. When petroleum products reach the end of their product lives, most are incinerated or thrown away. The plastic thrown into garbage dumps

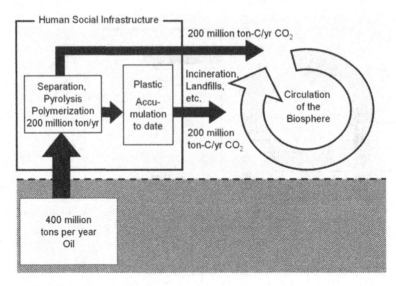

Fig. 1-8: The lifecycle of plastic

does not decompose quickly, but after a long time, it will eventually be oxidized into CO_2.

Consequently, seen from a global perspective, the lifecycle of plastic is just the transformation of oil from oil fields into CO_2 released into the atmosphere (figure 1-8).

Biomass Materials Are Recirculated

Iron, aluminum, concrete, glass and plastic have lifecycles that currently proceed in what is essentially a one-way flow from natural resources to release back to the environment as waste material and CO_2. In contrast, biomass is an example of a basic material that, in some cases, is recirculated even now.

Biological resources that are not used as food, such as wood and the husks of plants, are referred to as "biomass." Biomass materials include paper and lumber. Paper is made from trees; however, the process of making paper uses a rather large amount of fossil fuels. About half of a tree's wood consists of cellulose; the other half consists of lignin, a substance that keeps the trees rigid. Paper mills use only the cellulose to make paper. However, the lignin is not just thrown away; it is used as a fuel to generate electricity. Unfortunately, there is not enough lignin to supply all of the electricity required for paper production, so oil is used to cover the deficit. The worldwide production of paper is about 400 million tons per year, and about 300 kg of oil is used to make one ton of paper.

Currently, the recycle ratio for paper in Japan is about 50%. Although the recycle ratio varies from country to country, we can estimate that on average about half of the paper used in the world is recycled. Therefore, about half of the 400 million tons of paper produced per year is made from used paper. The rest of the used paper is either incinerated or thrown away in garbage dumps, where it is decomposed, oxidized, and finally becomes CO_2.

In summary, the lifecycle of paper begins with the harvesting of trees as raw material, and after the paper is used twice on average, it is released into the atmosphere as CO_2. The trees harvested to produce paper grow by acquiring CO_2 from the atmosphere. If the forests cut down to make paper are not replanted, the cycle of biomass material is not complete and the flow is one-way, like the flow for glass and cement. However, if the same number of trees that is harvested is replanted, the lifecycle proceeds from trees to paper to CO_2 and back to trees. This is essentially the same as the natural circulation of trees growing, dying, and decomposing. Therefore, biomass has a recirculating lifecycle that can be sustained in the biosphere. When we look at the overall lifecycle of paper produced this way, we see that the chief impact on the biosphere comes from the 300 kg of fossil fuels consumed per ton of paper, oil taken from the oil fields and released as CO_2 in the atmosphere (figure 1-9).

How about the lifecycle of lumber, the other major biomass material? If a wooden house is torn down at the end of its life and the wood is thrown into a garbage dump, the lifecycle will be a one-way flow. However, in making lumber, fossil fuels are used only to harvest, transport, and shape the wood. These processes consume far less energy than separating lignin from wood to make paper. Therefore,

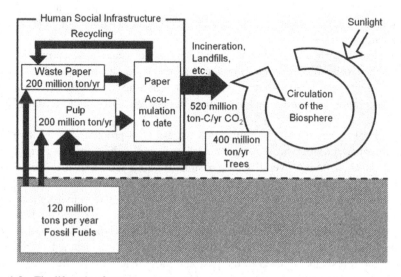

Fig. 1-9: The lifecycle of paper

as long as the trees that are cut down are replanted, consumption of lumber is sustainable. In essence, this lifecycle is the same as the circulation of biomass that occurred in nature before humans began to disturb it.

Sustainable Lifecycles and Non-sustainable Lifecycles

Looking from a global perspective at the processes for producing basic materials, we see that the consequences of "mass production / mass consumption" are quite different for different materials. The point to keep in mind is that it is possible to manufacture each of these materials in a sustainable way and a non-sustainable way. A large fraction of discarded iron and aluminum products is currently recovered as scrap and reused. But many metal products are still discarded without recycling. Most of the waste concrete produced when buildings and bridges are demolished is used for purposes such as road beds. However, as the demand for road bed and other low-grade materials decreases, the amount of concrete that is thrown away will increase. In some regions of the world, renewable forestry is practiced so that as trees are cut down, others are planted. But in other regions, forests are cut down without replanting, and the bare terrain is left to become a desert.

When materials are reused or resources replaced, the resources are not consumed in a one-way flow; instead they are circulated through human society twice or more. However, material flows that proceed directly from resource to waste should give us cause for alarm. For human activities to "fit" in the biosphere, they must circulate in the same way that natural biosphere activities do. Right now, we too often extract resources from the earth to make products and then return the discarded products to the earth, relying on the earth's natural circulations to complete the cycle back to resources. That is a "one-way" flow, and it has begun to overwhelm the capacity of the earth to stay in balance.

It is clear that our activities of "making things" are disrupting the natural circulations of the biosphere. However, those are not the only human activities threatening the earth. Our normal day-to-day activities such as driving cars, using air conditioning, and lighting our homes also have a great impact. We will call these "daily life" activities.

This book is based on the premise that the essential problem of sustainability is that human activities of "making things" and "daily life" are not carried out in accordance with any overall global vision. Without such a vision, we do not know what the future consequences of our present activities will be. In other words, we do not know whether activities touted as beneficial for the environment will actually result in the consequences we intend. This lack of a global vision is, I suggest, the reason for the widespread feeling of helplessness in regard to the sustainability of the earth. Human civilization has already consumed more than 40% of the forests that existed in the past and more than 50% of the recoverable oil resources. We cannot dismiss these numbers as groundless fears. We must, instead, find a way to marshal our efforts to achieve a sustainable earth. In the next section, you will see why.

5 What Happens if We Continue with "Business as Usual"?

Oil Reserves Will Become Depleted

Until this point, we have examined the present-day lifecycles of metals, cement and glass, plastics and paper – lifecycles driven by the human activities of manufacturing and consumption. If we continue with "business as usual," what will the earth be like by the middle of the 21st century?

We have seen that the production of all basic materials requires the combustion of large quantities of fossil fuels. To make one ton of plastic, we must burn one ton of oil. To make one ton of iron takes 600 kg of coal. We need 300 kg of fossil fuels to make one ton of paper, 200 kg to make one ton of glass, and 100 kg to make one ton of concrete. If we continue to use oil to provide the energy for manufacturing these materials, world oil reserves will almost certainly be depleted by the end of the 21st century.

It has been said that oil reserves will last at least another 40 years, but how is this number arrived at? The life expectancy of the world oil reserves is calculated as the total amount of confirmed reserves divided by the current annual consumption rate. Consequently, if new oil reserves are discovered and the amount of confirmed reserves is increased, the projected life expectancy will increase. On the other hand, if the annual consumption rate increases, the expected lifetime of the reserves will decrease. The reason that oil reserves have not yet been depleted, even though more than 40 years ago people were saying that oil reserves would only last 30 or 40 years, is that until now new oil fields have been discovered at a rate comparable to the rate of oil consumption.

However, the number of new oil fields discovered each year is decreasing, and the size of the newly discovered oil fields is getting smaller. In addition, more and more of the major existing oil fields are nearing the end of their reserves. For example, in the U.S., which in addition to being the largest oil consumer is the largest oil producer after Saudi Arabia and Russia, oil fields have already exceeded their peak output levels, and since the 1990's, the production rate there has been declining continuously. In 1998, the ratio of remaining reserves to annual production was less than ten years, and it was predicted at that time that, even with the discovery of new oil fields, after ten years the reserves would be almost completely depleted. According to more recent figures for 2006, the ratio of remaining reserves to annual production was still about ten years. However, the production rate declined by more than 20% from 1998 to 2006, despite the rise of world oil prices. This is a clear indication that the U.S. oil reserves are running out. The situation of the British oil fields in the North Sea is similar.

On the other hand, the rate of fossil fuel consumption worldwide continues to increase. It is a telling fact that China, home to one fourth of the world's population, changed from being an exporter to an importer for fossil fuels during first half of the 1990's. The increase in fossil fuel consumption resulting from the economic growth occurring in South East Asia is also remarkable. These changes in the world oil market all point towards the impending reality of oil depletion. In the past, human civilization has experienced two energy crises. However, those crises were caused more by political and economic factors driving up the price of oil, such as propagandistic reports that oil reserves might eventually be depleted, rather than real evidence that oil depletion could occur in the near future. Between 2050 and 2100, oil depletion may become a reality, leading to a different, more fundamental, sort of crisis.

Global Warming Will Alter the Earth's Climate

The second catastrophic event that is almost certain to occur in the 21st century is global warming. Despite the clear messages from authorities such as the IPCC, some people still claim that there is scientific uncertainty about global warming. But just looking at the mechanisms by which global warming occurs, it is clear that global warming is an undeniable reality.

The earth's surface temperature is sustained by heat from the sun. Without the sun, the earth would cool down to near the temperature of outer space, which is about $-270°C$. The reason that your hands get warm when you hold them up to a wood stove is that energy radiates from the hot stove and heats your hands. The higher the temperature of an object, the more energy radiates from its surface, mostly as infrared radiation, which we feel as heat. The energy radiating from the sun shines on the earth at a rate of approximately 1.4 kW per square meter, and this energy heats the earth. However, as shown in figure 1-10, energy is also released from the earth's surface into outer space in the form of infrared radiation. In fact, the temperature of the earth's surface is just high enough that it releases an amount of energy into space exactly equal to the energy arriving from the sun. If the temperature increases, the amount of infrared radiation leaving the earth increases, causing the temperature to fall. If the temperature decreases, the amount of infrared radiation becomes smaller causing the temperature to rise. Therefore, the earth's surface temperature is maintained by a balance of energy radiation. If the earth had no atmosphere, the balance temperature would be $5°C$.

The earth's atmosphere affects this balance temperature in two ways. The first effect comes from the clouds and particles in the atmosphere, which reflect part of the sunlight and keep it from reaching the earth's surface. The fraction of sunlight that is reflected is about 30%. This reflected sunlight reduces the balance temperature by $23°C$. Without the second effect of the atmosphere, that would result in a surface temperature on earth of $-18°C$.

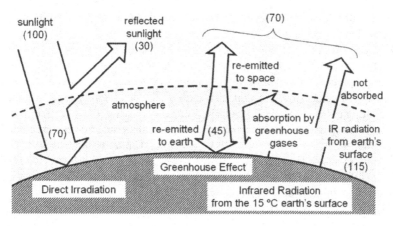

Fig. 1-10: The energy balance determining the temperature of the earth's surface

The second effect of the atmosphere is the absorption of infrared radiation emitted from the earth's surface by molecules of particular gases, such as water vapor and CO_2. These radiation-absorbing molecules are called "greenhouse gases" because they trap heat in the same way as the glass roofs of greenhouses. After molecules absorb infrared radiation moving from the surface of the earth towards outer space, they re-emit it immediately, but only half of the re-emitted radiation is released towards outer space. The other half is redirected back towards the earth's surface. Some of the infrared radiation released towards outer space is reabsorbed by molecules in the atmosphere still further from the earth's surface, and half of that radiation is re-released in the direction of the earth's surface.

This absorption and emission continues until the radiation is returned to the earth's surface or escapes into outer space. The result is that an amount of infrared radiation equivalent to more than 60% of the sunlight that reaches the earth's surface is captured by the atmosphere and returned to the earth's surface. This is the greenhouse effect, and it currently raises the temperature of the earth's surface by about 33°C.

The net result of a 23°C temperature decrease from reflection and a 33°C temperature increase from the greenhouse effect is an increase in 10°C, which when added to the 5°C temperature of the earth without its atmosphere gives us the actual average temperature of the earth's surface: 15°C.

Venus, the planet next to the earth in proximity to the sun, has a surface temperature of 400°C, and Mars, the planet next to the earth in distance from the sun, has a surface temperature of −50°C. Both these temperatures are determined by the same mechanisms that regulate the earth's temperature. Therefore, if the concentration of a greenhouse gas such as CO_2 increases, it is reasonable to conclude that the greenhouse effect will increase. Scientists predict that the rise in the earth's average temperature when the concentration of CO_2 doubles will be 3°C. Looking back at the rate of increase in CO_2 concentration shown in figure 1-1, it is clear that by the middle of the 21st century a significant increase in global warming from CO_2 emissions is inevitable.

No one knows for sure what the effects on the earth and on human civilization will be from such an increase in global warming. However, we do know that it will mark an unprecedented change in the history of human civilization. Certainly, fundamental changes will occur in the earth's climate, such as rainfall patterns, with resulting effects on crop productivity. As we saw earlier, the level of the oceans is already rising, and there is reason to believe that the rise could be large enough to cause significant parts of the world's coastlines and entire island nations to disappear beneath the sea. If we continue with business as usual, it is almost certain that by the middle of the 21st century the earth's energy balance will require us to make major changes in the way we live.

The Earth Will Become Buried in Human Waste

The third crisis that we will face by the middle of the 21st century is the accumulation of massive amounts of waste material.

As we saw in figure 1-4, of all human artifacts existing in society today, most of them were produced in the latter half of the 20th century, and there is no sign of decline in the rate of production. These human artifacts accumulate mainly in cities, where the greatest population increases have occurred in the 20th century. And people are continuing to migrate to the cities, particularly in developing countries. It is predicted that by the middle of the 21st century, 70% of the world population will be living in cities. As existing cities expand and new cities are built, the accumulations of human artifacts will also grow. However, all things must reach an end. The life span for products such as automobiles and household appliances is about 10 years, and for buildings it is around 40 to 50 years. Therefore, almost all of the human artifacts that we see in the cities today will reach the end of their life spans by the middle of the 21st century. When the mountains of human artifacts accumulated in the second half of the 20th century reach the end of their product lives, a massive generation of waste materials like nothing we have seen before will begin. If this waste material is thrown away as garbage, dumps will have to be created all over the surface of the earth to hold it all.

Around the world, it is becoming difficult to obtain sites for garbage dumps. Intense debates have sprung up when plans to create garbage dumps are announced that involve destroying fragile ecosystems such as tidal wetlands. On the other hand, illegal dumping of garbage has become conspicuous on islands of the *Seto Inland Sea*, in suburbs of major cities, and in forestlands everywhere. And this is just the beginning.

These phenomena – depletion of oil, global warming, and the massive generation of waste – are natural results of the explosive expansion of human activities in the 20th century. And it is under these severe circumstances that we enter the 21st century.

Powered by the energy of the sun, the earth has maintained the various cycles of nature within the thin layer of the biosphere since before human civilization began. Now human activities are threatening to disrupt these cycles. To achieve a sustainable earth, it is up to us to figure out how to construct a sustainable circulation system for our own activities, a system fits within the natural circulations of the earth. The purpose of this book is to show that this can be done.

Chapter 2
Knowing Energy

Any action that does not happen naturally or spontaneously, such as lifting some-thing heavy from a low place to a high place or moving heat from a cold place to a hot place, requires energy. Because almost none of the human activities of "making things" and "daily life" occur spontaneously, they nearly all require energy. Therefore, energy is an essential piece of the puzzle in figuring out how to sustain the biosphere while we provide a modern standard of living for the human population of the earth. But many fundamental concepts of energy are difficult to grasp. Although a lesson on what energy is and what it means to consume energy may seem unexpected in a book about creating a sustainable society, it is important that we clarify these concepts before introducing ways in which technology can be used to make human existence on the earth sustainable.

1 Energy Is Conserved

Energy Is the Ability to Do "Work"

You have probably seen a building demolition team use a crane to lift an iron ball and drop it to break up concrete structures. When any object, not just an iron ball, is dropped from a high place, it can do "work." "Work," like energy, is a word we use in many ways in ordinary conversation; however, in the world of science and technology, "work" has a strict definition. "Work" is defined as the product of a force and the distance that an object is moved by applying that force. For example, when an iron ball is raised a certain distance, the "work" done equals the force applied to the ball times the distance the ball is raised. To raise the ball twice the distance, twice as much work is required, and if the weight of the ball is reduced to half, then half the work is enough to raise the ball. However, work can also take the form of other changes. For example, crushing concrete structures is a form of work that is done by the iron ball dropped from the crane. An iron ball flung through

Hiroshi Komiyama and Steven Kraines
Vision 2050: Roadmap for a Sustainable Earth.
© Springer 2008

the air does work when it hits a thin sheet of iron and changes the shape of the iron sheet. A definition of energy that is appropriate for the discussion in this book is the ability of physical objects and their conditions, such as their temperatures and pressures, to do work.

Kinds of Energy

There are three basic types of energy: external energy, internal energy, and field energy. The energy contained in the iron ball that is lifted up by the crane is called potential energy, and the energy of the ball flying through the air is called kinetic energy; these are both types of external energy. Other objects that have potential energy include helicopters hovering in the air, water held up in a dam, and a car stopped at the top of a hill. Other objects having kinetic energy include a moving car, flowing air, and a spinning motor.

"External energy," such as the motion of a car or the position of a helicopter high above the ground, is apparent from outside. In contrast, "internal energy" is energy hidden in the object itself, which cannot be detected from outside appearances. Internal energy includes heat energy, chemical energy, and nuclear energy.

Although most people have a vague understanding of what "heat energy" is, the term "heat energy" is not actually correct. It is more accurate to say that objects having a high temperature have internal energy. Recall that the definition of energy is the ability to do work. If we have water with a high temperature, then we can use it to make steam and use the resulting pressure of the steam to drive a steam engine, making it possible, for example, to do the work of moving a steam locomotive. In other words, water with a high temperature has internal energy that can be used to do work in the same way as kinetic and potential energy. "Heat," on the other hand, is the transport of internal energy from an object having a high temperature to an object having a lower temperature.

Fuels such as kerosene have internal energy in the form of chemical energy. If we combine kerosene with oxygen in a chemical reaction called combustion, a hot flame will be produced, a flame we can use to turn water into steam. Nuclear energy is contained in every atom; however, there are only a few elements whose atoms can be easily used to obtain energy for doing work. One of those elements is uranium. We can use the heat generated when an atom of uranium is split into smaller atoms through the process of nuclear fission to do work, for example to make electricity. Nuclear fusion is another process that creates heat from nuclear energy. When two hydrogen nuclei are fused together to make a helium nucleus, heat is also released. This fusion is what powers the sun.

Field energy can be imagined by thinking of the inside of a microwave oven. When you turn a microwave oven on, the inside becomes filled with electromagnetic waves, which is a form of "field energy," and that energy can do the work of raising the temperature of the cup of coffee in front of you that has gotten cold as you were reading this chapter.

Energy Media and the Law of Energy Conservation

We often refer to work, heat, electricity and light as "energy;" however, strictly speaking, they are energy media, that is, ways for transporting energy from one object to another. For example, if we burn some propane to heat the water in a teapot, the chemical energy that was in the propane is changed into the internal energy of the water through the medium of heat, resulting in the rise of the water's temperature.

The energy of an object can be used to do work, and work can be used to add energy to an object. Think back to the iron ball being dropped from the crane to break up a building. When the crane lifts the iron ball, the ball will gain no more potential energy than the amount of work that is applied to it by the crane. When the iron ball is released from some height, it will fall. As it falls, it loses potential energy corresponding to the distance that it has fallen, and its kinetic energy increases by essentially the same amount. So as the ball falls, potential energy is transformed into kinetic energy. The form of the energy is transformed, but the total amount of energy – the sum of the potential energy and the kinetic energy – remains constant, as illustrated in figure 2-1. As a general principle, when energy changes from one form to another, the total amount of energy in all forms remains the same. This principle is called the law of energy conservation.

What happens when the iron ball hits the ground and stops? Both the kinetic energy and the potential energy of the ball are gone because the ball is no longer at a high location, nor is it moving. But if that is all that happens, the law of energy conservation will be abrogated. Actually, when the iron ball hits the ground, heat is generated, and the temperature of the ground and the surrounding air is raised. When we say that the temperature rises, we mean that the internal energy of the ground and air is increased, and this increase in internal energy is exactly the same as the potential energy of the iron ball before it fell. Furthermore, this amount of

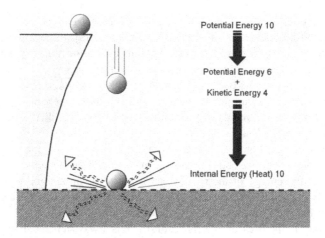

Fig. 2-1: Transformation and conservation of energy

energy is also the same as the work that is required to lift the iron ball back to its original height.

In this way, the forms of energy can be changed, but the total amount is conserved. But if that is the case, how can we talk about an energy crisis or say that a form of energy is being depleted? We shall return to this question in the third section of this chapter.

Because our explanation of energy has been brief and because the concepts can be a bit tricky to grasp, let's pose a few questions here that may bring these concepts of energy closer to home.

Question 1: In a closed room, which has a larger heating effect: turning on a 1 kW electric heater, or turning on television sets, radios and lights with a total power rating of 1 kW?

Answer:

A) turning on a 1 kW electric heater
B) turning on the televisions and other appliances
C) almost the same
D) exactly the same

Question 2: If you leave the door of a refrigerator open in a closed room, what will happen to the room's temperature?

Answer:

A) the temperature will increase
B) the temperature will decrease
C) the temperature will not change much
D) the temperature will not change at all

The answer to question 1 is "almost the same." After electricity is transformed into light and sound by television sets, radios, and lights, all of the energy in the end becomes heat, so the heating effect of the appliances is almost the same as turning on a 1 kW heater. The reason that the answer is "almost the same" is that since we can see light from the television and hear sound from the radio from outside the room, we know that a small part of the energy from those appliances escapes the room through the energy media of light and sound. Therefore, there will be a very small difference in the heating effect.

The answer to question 2 is "the temperature will rise." This may seem counterintuitive to you, but if we consider the law of conservation of energy in the closed room, the internal energy of the room must increase by an amount equivalent to the electricity consumed by the refrigerator. A refrigerator is actually just a device for pumping out the heat that leaks into the space inside the refrigerator from the air in the room. In the back of a refrigerator, there is always a place that is hot, and from that place heat is released into the room. If the refrigerator door is left open, the amount of heat released from the back of the refrigerator will be more than the cooling effect coming from the open door. The difference is exactly the amount of electricity that is consumed. Recently, in places like hotels, refrigerators are often placed in a box made to look like a piece of furniture in order to keep them out of sight. However, if there are not enough openings in the box, it will get hotter and hotter until the refrigerator ceases to work. Many of you who travel a lot have probably stayed in hotel rooms having this problem.

Here is one more question (the last, I assure you!).

Question 3: In the situations described in the previous two questions, where does the heat generated from the electricity go?

If every time we use energy, that energy ends up warming the surrounding air as heat, why is it that the temperature of the earth does not rise? The reason is that in the end, energy that becomes an increase in the temperature of the air and the surrounding environment, what is called the "ambient temperature," is finally radiated to outer space as infrared radiation. As we saw in the section describing the mechanism of global warming from the previous chapter, when the temperature of the earth's surface starts to rise, the radiation from the earth increases, thereby keeping the temperature stable. An increase in radiation from the earth means that energy that has taken the form of an increase in ambient temperature is escaping to outer space through the medium of heat.

How a Thermal Power Plant Works

Among the many different media for energy, electricity is one of the most outstanding. Electricity can be easily changed into light, work, or heat; it can be transported using just a wire, and it can be turned on and off with a single switch. The amount of energy a nation consumes usually increases with improvement in living standards, and the increase is especially large for electricity. However, unlike forms of energy such as gasoline, which we can see, electricity is invisible, so it can be more difficult to understand. Let's summarize the main concepts here. There are two methods for obtaining electricity. One is to use an electric generator. The other is to use an electric cell.

An electric generator works the same way as a generator-type light on a bicycle. You know those non-battery powered bicycle lights with a little wheel that is turned by the front wheel of your bicycle? These generator-type lights contain a magnet that is placed around a coil of metal wire. The coil can be turned on an axis, and when it rotates inside the magnet, electricity is generated that flows through the coil. Therefore, an electric generator is a mechanism for transforming rotational work into electricity. There are many techniques used to rotate the coil. In the case of the bicycle light, the rotation energy comes from the wheel that turns when you pedal. In wind power, wind is used to create rotation energy by turning the blades of a wind turbine. In hydropower, the force of water that flows down through a pipe turns the blades of an impeller.

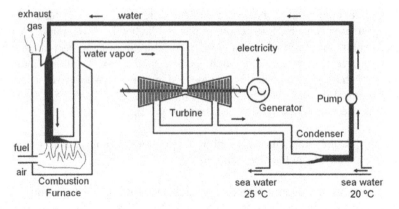

Fig. 2-2: The basic mechanism of thermal electric power generation

Fig. 2-3: The turbine of a thermal power plant (Courtesy of Tokyo Electric Power Company)

Figure 2-2 shows a conceptual image of the mechanism of a thermal power plant. First, fuel is combusted in a furnace and used to boil water in steel pipes, producing steam. Then, the steam is channeled to a turbine, causing it to turn and thus producing rotational energy that is transformed into electricity using a generator. Figure 2-3 is a picture of a turbine with its outer cover removed. A turbine is basically a huge high-precision wind mill made of a special kind of steel that is rotated using the force of steam. However, if the exit of the turbine is not at a low pressure, the steam will not flow through the turbine. Therefore, the exit is connected to a steam condenser made of numerous thin pipes through which water or some other coolant flows. By changing the steam to water in the condenser, the pressure is reduced causing more steam to flow through the turbine. The condensed water is returned to the furnace using a pump. In short, water is circulated from the combustion furnace, and during that circulation it turns the turbine which drives the electric generator. In this way, we are able to extract electricity from the chemical energy of fuel. However, less than half of the chemical energy of the fuel can actually be transformed into electricity. Most of the heat produced by combustion of fuel is lost when the steam is condensed in the condenser. As a result, more than half of the chemical energy of the fuel used in a thermal power plant is released as waste heat into the environment.

The mechanism of a nuclear power plant is essentially the same as that of a thermal power plant. The main difference is that in place of the furnace where fuel is combusted in a thermal power plant, a nuclear power plant uses a nuclear reactor, which produces heat from nuclear fission.

How Electric Cells Work

There are many kinds of electric cells. Chemical electric cells change chemical energy into electricity. Solar electric cells, which are usually just called solar cells, change sunlight into electricity. We will see how solar cells work in Chapter 6. Currently, most of the widely used chemical electric cells work by separating two chemicals with a fluid or some kind of separating membrane that is porous only to ions, placing electrodes in each chemical, and allowing the two chemicals to react.

You may recall from high school chemistry experiments that water molecules can be separated into hydrogen and oxygen by applying electricity. This is called the electrolysis of water. As shown in figure 2-4, a hydrogen-oxygen chemical electric cell uses the same mechanism, except that at the places where hydrogen and oxygen are produced in electrolysis of water, hydrogen and oxygen are supplied in a chemical electric cell, and at the place where electricity is provided in electrolysis of water, electricity is extracted in a chemical electric cell. Here is how the chemical electric cell in figure 2-4 works. The membrane of the cell is made of a material that allows only hydrogen ions to pass through. Thus, the only way for hydrogen on the left side of the membrane to get over to the right side so that it can react with oxygen to produce water is for the molecules of hydrogen to give up electrons and change into hydrogen ions. Once the atoms in the hydrogen molecules are changed into ions, they can pass through the membrane to the right side of the cell, but the electrons cannot. The electrons are needed to complete the reaction of oxygen and hydrogen to water, so they must find another way to get to the right side of the cell. This way is provided by an external circuit that connects electrodes on each side of the cell. The electrons travel via the external circuit to reach the right side of the cell, where they change the oxygen molecules into oxygen ions. The oxygen ions then react with the hydrogen ions that passed through the membrane, thereby forming water. In this process of making water from hydrogen

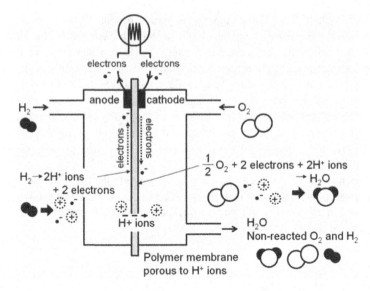

Fig. 2-4: The basic mechanism of an electric cell
Note: Ions are formed on both sides of a membrane that prevents the passage of electrons. The
ions on one side pass through the membrane to react with the ions on the other side. The electrons
travel through an external circuit and become electricity. The example in the figure using hydrogen
and oxygen is called a fuel cell.

and oxygen, electricity can be extracted in the form of the flow of electrons through
the external circuit.

In essence, hydrogen and oxygen have a natural tendency to combine spontaneously and form water, and that natural tendency can be harnessed to produce electricity. This is a specific example of the general rule that any chemical process that
proceeds spontaneously can produce work.

There are many kinds of chemical electric cells. Each kind of cell has a different
combination of the reacting chemicals involved and the membrane or other separator
used to separate the chemicals. The most common chemical electric cell, the dry
cell, uses magnesium dioxide and zinc. Lithium batteries use magnesium dioxide
and lithium separated by a thin sheet of plastic, mercury batteries use mercury oxide
and zinc, and car batteries use lead oxide and lead separated by sulfuric acid. The
electric cell in figure 2-4, which uses hydrogen and oxygen, is called a fuel cell.

Energy Resources

When experts talk about energy crises, they are referring to the problem of a depletion or inadequate distribution of energy resources. So what is an energy resource?
Basically, an energy resource is a source from which or a method by which energy

can be obtained. However, when we speak of energy resources in the context of the sustainability of the earth, what we usually mean is "natural energy resources," or sources of energy obtained directly from nature. Natural energy resources may be buried in the earth, growing on the earth's surface, or falling from the sky. However, no artificial processes are necessary to create these resources.

Hydrogen and electricity are not energy resources. The reason is that, for all practical purposes, these sources of energy cannot be obtained directly from nature. There are few people who consider electricity to be an energy resource; however, strangely, many people misunderstand hydrogen. It is often said that "hydrogen can solve the energy problem" or that "we can create a country based on hydrogen." The gist of these claims is that, because it is possible to make hydrogen from the electrolysis of water, and there is an abundant supply of water, if we were to use hydrogen to meet our energy needs, we could solve the energy problem and simultaneously end the emission of toxic materials. But this is not correct. Even if there were an inexhaustible supply of water, electricity is required to obtain hydrogen from water, which puts us back in the position of needing an energy resource to produce the electricity. To use hydrogen as a source of energy, we still must draw on some energy resource to obtain the hydrogen.

Therefore, in addition to fossil fuels and nuclear energy, the energy resources that we know about consist of geothermal energy (which is the energy of the earth's core), the rise and fall of the tides (which are pulled by the moon), and solar energy, including all of the energy resources powered by the sun, such as wind, rain, and biomass. Currently, almost 80% of the energy used worldwide is supplied by fossil fuels, including oil, coal and natural gas. Solar energy in the form of biomass and hydropower supplies about 15%, and nuclear energy supplies about 5%. Geothermal energy, tidal power, and forms of solar energy other than biomass or hydropower together make up less than 1%. The role of oil refineries and power plants is to transform energy resources into forms that are easy to transport and easy to use, such as gasoline, compressed or liquefied natural gas, and electricity. The role of engines, motors, appliances, and lighting fixtures is to transform these forms of energy into the work, heat, and light that we use directly in "making things" and "daily life."

Expressions for Energy

There are several methods to express measures of energy resources. "Coal conversion" and "oil conversion" are methods whereby a form of coal or oil is chosen, and its heating value per unit mass is taken as a standard unit. Then quantities of other energy resources needed to do a given amount of work are converted into those standard units. The numbers in figure 2-5, which we will see in the next section, use a form of oil conversion called TOE, for ton-oil-equivalent. There are methods for expressing nuclear energy and hydro energy in the same way. Because most energy resources are used as electricity, we need a way to express how many

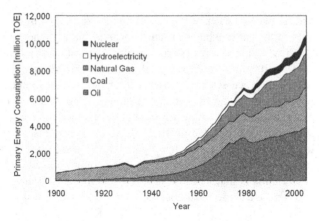

Fig. 2-5: Global consumption of energy from 1900 to 2008 (Data from the BP Statistical Review of World Energy 2007)

standard "oil conversion" or "coal conversion" units a given amount of electricity corresponds to.

There are two ways to do this. One way is to calculate the amount of heat that can be produced by using the electricity in an electric heater. Then, this amount of electricity is expressed in terms of the amount of fossil fuel needed to provide the same amount of heat. The other way is to calculate the amount of heat needed to produce a given amount of electricity in a thermal power plant. Earlier, we saw that less than half the heat energy of the fossil fuel consumed is actually transformed into electricity. The amount of fossil fuel needed to produce some amount of electricity is that amount of electricity divided by the generation efficiency of the power plant. If the generation efficiency is 33%, then three times as much fossil fuel energy is required. Using this second method gives a more accurate assessment of how much fossil fuel would be required to meet some energy demand if all of the energy were provided by fossil fuels. But if we convert the electricity produced by nuclear power and hydro power into standard units using the first method, we will underestimate the amount of nuclear and hydropower energy used. The amount of primary energy consumption provided by nuclear power plants and hydropower plants in figure 2-5 is obtained by dividing the electricity provided by the plants by a power generation efficiency of 0.33, the global average for thermal power plants.

Another way to express the measure of an energy resource is by converting to units of carbon. In this method, each energy resource is expressed as the amount of carbon contained in the resource. Therefore, this method is applicable only to carbon-based fuels and cannot be used for energy resources such as nuclear and hydropower. And this method cannot accurately compare energy resources that yield large heating values per unit of carbon, resources like natural gas, with energy resources like coal that are highly carbon intensive. Nevertheless, because the global warming is basically caused by CO_2, we can, by converting fossil fuel resources into carbon units, directly express the effect of burning those resources on global warming. In this book, when referring to precise values of energy amounts, we will use oil conversion units; and in all other cases, we will use carbon conversion units.

2 What Is Energy Used for?

World Energy Consumption Is One Ton per Person per Year

In figure 2-5, we see how dramatic the rise in energy consumption has been in the 20th century, an increase of approximately 20-fold. Today, the amount of fossil fuel consumed annually (about 80% of the total energy consumption) is about 7.5 billion tons when converted to carbon units. Because the current world population is more than 6.5 billion people, the average consumption of fossil fuel energy by the people of the world at the turn of the century was just a little more than one ton per person per year.

So how do the numbers look in Japan? Japan has a population about 125 million and consumes about 350 million tons of fossil fuels, so it has a per capita fossil fuel consumption of 2.7 tons. Almost all of the fossil fuels imported to Japan each year are first sent to oil refineries, electric power plants, and gas companies. Currently, the distribution is 60% for oil refineries, 25% for power plants, and 5% for gas companies. The remainder of the fossil fuel is coal used for making iron and steel. The oil refineries, power plants and gas companies do not use the energy themselves but instead deliver it to places where it is needed in the activities of "daily life" and "making things."

So how is all of this energy used? To answer this question, we would need data on the distribution of energy use for all of the countries in the world. Unfortunately, such information is not generally available, even in many developed countries. Japan is one of the few countries that has data on the distributions of energy use, so we will illustrate the concepts of energy use for "daily life" activities with data from Japan.

The distribution of energy use in Japan is shown in figure 2-6. The places where "daily life" activities occur are homes, offices and transportation, accounting for 9.5%, 13.0% and 16.5%, or a total of 39% of the energy consumption. Industry, that is "making things," consumes 31%, and 30% is consumed in transforming various forms of energy into electricity and oil refinery products. Next, let us examine how energy is used in each of these activities.

Energy Use in "Daily Life"

Energy is consumed through "daily life" activities in homes, work places, and transportation. The energy consumed in homes consists almost entirely of electricity,

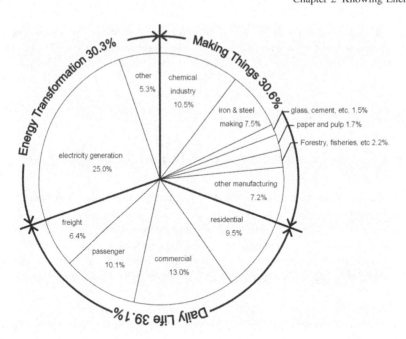

Fig. 2-6: Distribution of energy consumption in Japan (Data from Sogo Enerugi Tokei 2005, Japan Agency for National Resources and Energy)
Notes: Data is from 2007. The energy consumed in "energy transformation" is mostly energy in power plants that does not become electricity or that is used internally in the plant.

gas, and kerosene. This energy is used to cook, heat water, run electrical appliances, and heat or cool the home. Energy consumption in offices and other work places is not much different, although there is some variation in the way energy use is distributed, with a greater consumption of electricity in the work place by computers and copy machines.

Refrigerators, washing machines, and vacuum cleaners all work by using electricity to drive a motor. These uses of energy, together with lighting and televisions, make up about one third of the total consumption of energy from "daily life" activities in homes. Roughly speaking, another third is used for heating and cooling, and the last third is used for hot water and food preparation. Of the total household use of electricity, refrigerators, air conditioners, and lights each consumes about 20%.

Forms of transportation that use energy include passenger cars, trucks, buses, trains, airplanes, ships and so on. In Japan, gasoline for passenger cars accounts for more than 50% of the total energy consumption for transportation, both personal and business related. The next largest contribution is the 35% used by freight vehicles for business and personal transport, mainly trucks. Other forms of transport, such as planes, ships, taxis, buses, and trains, constitute less than 15%. Therefore, even if we assume the energy used in transportation to be just the amount used in cars and trucks, our error will not be so great.

The Production of Basic Materials Is the Core
of Manufacturing

We can readily picture how energy is consumed in "daily life;" however, the consumption of energy in manufacturing may be somewhat more difficult to imagine. The manufacturing process that consumes the most energy is the making of iron and steel, followed in order by the production of chemical materials like plastics, non-metal minerals like glass and cement, and paper and pulp. In Japan, these industries alone account for more than 60% of the energy consumed in manufacturing. That is, most of the energy consumption in manufacturing is used to change natural resources into basic manufacturing materials such as iron, cement, glass, paper, plastic, synthetic fibers and rubber. As we saw before, the quantity of fossil fuels needed to make one ton of material is 600 kg for iron, one ton for plastic, 100 kg for cement, 200 kg for glass, and 300 kg for paper. This is the nature of energy consumption in manufacturing. The combustion of fossil fuels in the global flow of basic materials, the flow we looked at in the previous chapter, accounts for nearly all the energy consumed in "making things."

You may have noticed that in the list of industries consuming the most energy, the manufacturers of cars, heavy equipment, and home appliances are not included. Construction and urban engineering companies are also missing. The reason is that, in comparison to the energy used in producing basic materials, very little energy is consumed at assembly plants and construction sites.

Consider the example of a car. The largest energy consumption in a car's lifetime is the gasoline used to drive it. The next largest is the energy used to produce the basic materials of the car, such as iron and plastic. These materials are purchased by automobile companies and assembled into cars; however, the energy consumed by shaping the materials and assembling them is surprisingly small. According to one estimate, of the total energy consumed by a car – from production to disposal – 79% goes to the gasoline use to drive it and 14.5% to basic materials used to make it. Only 4.5% goes to the process of assembling it, with the remaining 2% used for maintenance, repair, and disposal.

We often see giant cranes at construction sites with sparks flying as workers solder parts together, and on the television, we see video footage of factories using robots and conveyer belts in assembly lines. But the amount of energy consumed at these stages of "making things" is surprisingly small. In fact, to determine which products consume the most energy in their manufacture, instead laboriously totaling the amounts of energy that different industries use to operate their machines and facilities, it is easier and almost as accurate to compare the energy consumed to produce the basic materials used to make the products. For example, in Japan about 50% of the iron produced is used in the construction of buildings and bridges, and 16%, in making automobiles. Thus, we can estimate that constructing buildings and bridges consumes about three times as much energy as manufacturing automobiles. Basic materials are produced to make the things that we consumers use, and it is in producing basic materials that the bulk of fossil fuels in manufacturing are consumed.

Energy Loss in the Energy Conversion Sector

Power plants, oil refineries, and gas companies are the main players in the energy conversion sector. The purpose of this sector is to change energy into forms that are easy for consumers to use. But it is never possible to convert 100% of one form of useful energy such as work into another such as electricity. During any transformation of energy from one useful form to another, some energy will always be transformed into heat at ambient temperature, which cannot be used. As a result, some part of the energy resources is consumed in the energy conversion sector. We saw earlier how thermal power plants fired by fossil fuels release over half of the fuel's chemical energy into the sea or atmosphere. In addition to that, a percentage of the generated electricity is consumed in operating the electric power plant itself. In the case of nuclear power plants, the power generation efficiency is lower, resulting in an energy loss of about 70%.

The fraction of electricity consumed in the operation of electric power plants around the world varies according to a number of factors, including the efficiency of the plant's operation and the technologies used to control pollution. For example, in Japanese fossil-fuel fired power plants, the ratio of electricity consumed by the plant itself is relatively high because almost all Japanese power plants use energy-consuming processes to remove sulfur oxides, nitrogen oxides, and fly ash from the combustion gas. As of 1990, world-wide there were about 2360 plants operating desulfurization equipment and 490 plants with denitrification, of which 1800 of the desulfurization plants and 350 of the denitrification plants were in Japan. Japan, a country that accounts for no more than 5% of the world's energy consumption and has no more of 5% of the world's power plants, operates more than 70% of the world's power plants with facilities for treating combustion gas. Thus it seems fair to say that in 1990 the only country doing a substantial amount of desulfurization and denitrification at power plants is Japan.

Obviously, by removing all of the desulfurization and denitrification equipment, we could increase the efficiency of fossil-fuel fired power plants. But it is hardly a reasonable solution. We must be vigilant to avoid approaches that increase efficiency only by creating other kinds of problems.

Since the 1990's, how much have other countries cleaned up their power plants? Figure 2-7 shows how much sulfur oxides were emitted on average per unit of electricity generated in 1999 and 2002 from thermal power plants using fossil fuels in several different countries. Most of the countries shown have decreased their sulfuric oxide emissions, and Germany now emits less than one gram of sulfuric oxides per kilowatt hour of electricity. However, even in Germany, fossil-fuel fired plants still emit more than three times the pollution of Japanese plants. When sulfuric oxides and nitrogen oxides dissolve in water, they become sulfuric acid and nitric acid, the precursors of acid rain. So it should be no surprise that the effects of acid rain on ecosystems are more serious in America and Europe than in Japan, although recently acid rain from China and other rapidly industrializing countries in East Asia is becoming a serious problem in Japan.

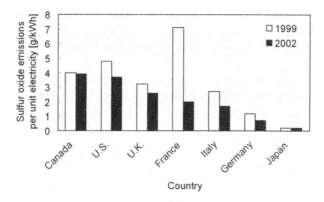

Fig. 2-7: Comparison of sulfur oxide emissions from thermal power plants (Data from Tokyo Electric Power Company)

The role of oil refineries is to separate oil into products such as gasoline, kerosene, light oil, and heavy oil, and then distribute those products to the places where they are used. The proportion of demand for the different component products of oil has varied by era as well as by country. For example, in Japan, during the era of fast economic growth following the Second World War, which centered on heavy industry and chemical plants, there was a large demand for heavy oil. At that time, about half of oil imports were refined into heavy oil. However, following that era, as a result of industrial advances in energy conservation and the increased use of automobiles, the relative demand for gasoline has increased. And now over 25% percent of imported oil is made into gasoline.

Oil refineries accommodate these changes in demand by adjusting the proportion of components in the final product. Like all other fossil fuels, oil consists mainly of carbon and hydrogen. Different refinery products have different ratios of carbon and hydrogen. Heavy oil, for example, has more carbon whereas gasoline has more hydrogen. As a result of the shift in demand from heavy oil to gasoline, most refinery products now contain more hydrogen than crude oil does. To increase the proportion of hydrogen, part of the oil is combusted, and with the energy produced, hydrogen is extracted from water and added to the oil. Using the added hydrogen, the amount of gasoline can be increased. Oil refineries must consume energy to carry out this process. This consumption rate is generally expressed as the fraction of the chemical energy of the crude oil entering the refinery that is retained in the chemical energy of the products. For modern day refineries, this fraction is about 95%. In other words, about 5% of the oil that passes through an oil refinery is consumed in the refinery process. This loss is much smaller than the loss of fuel energy in electric power plants, so making fuels like gasoline and kerosene is much less energy-intensive than making electricity.

The current role of gas companies is simply to distribute natural gas to consumers. However, if cogeneration systems can be made more efficient, for example through the application of fuel cells, gas companies could play an important role in spreading this technology by providing the necessary fuel supply networks.

Figure 2-6 shows the energy consumed for "daily life" and "making things" in Japan together with the amount consumed during energy conversion as described above. Nearly all of the energy consumed in energy conversion is the heat lost to the environment during the generation of electricity. The energy that has been converted into easy-to-use forms such as electricity, gasoline, kerosene and city gas is used for "daily life" and "making things," each of which consumes approximately half of that energy. Hence, we see the many ways in which energy is consumed in human activities leading to the supply of the products and services we use each day.

3 Energy Degrades

What Is the Value of Energy?

According to the law of energy conservation, energy is conserved. That is, the amount of energy before and after a change is always the same. However, the value of electricity, which can be used to turn on a television or run a vacuum cleaner, is totally different from the value of ambient heat, which is heat at the temperature of the environment, even if the amount of energy is the same. In other words, the value of energy is determined not only by its amount but also by its usefulness.

Under ideal conditions, a high quality motor can convert almost 100% of energy in the form of electricity into work. Similarly, a high quality electric generator can convert almost 100% of energy in the form of work into electricity. So intuitively it is clear that electricity and work have the same value. Furthermore, it is possible to convert nearly 100% of kinetic energy and potential energy into work. In short, work, kinetic energy, potential energy, electricity and all other kinds of energy except heat can be considered to have the same value.

Electric utilities exploit this property of energy by pumping water upstream of a hydropower dam to store electricity. In most developed countries, the demand for electricity is greater during the day than at night. However, for many forms of power generation such as nuclear power, it is not possible to stop plant operation at night and restart it in the morning, so a surplus of electricity is produced at night. At that time, water downstream of the dam of a hydropower plant is pumped up and stored in the upstream reservoir. The next day, when the demand for electricity is large, that water is let through the dam to generate electricity. The water is pumped up using a motor driven by the excess electricity generated by the nuclear power plant at night, which adds potential energy to the water. That potential energy is converted back to electricity when the water is released through the dam again. The ideal transformation efficiencies for these processes are all 100%, so it should be possible to retrieve 100% of the nighttime electricity produced by the nuclear power plant for supplying electricity in the daytime from the hydropower plant. But in reality 70% is the best that can be achieved today.

A 100% conversion between different forms of energy, which would be possible under ideal conditions, cannot be achieved in reality because every time energy is transformed, a part of the energy becomes heat. The reason electrical appliances – such as televisions, vacuum cleaners and light bulbs – become hot when we use them is that during the process by which electricity is transformed into light, sound, kinetic energy and so on, a part of the electricity is converted to heat. This occurs regardless of whether the device is used to produce light, sound, motion, or any other useful service. However, the fraction of electricity that becomes heat in different devices varies dramatically. In a hydropower electric plant, which is an example of a highly efficient system, about 85% of the potential energy of the water behind the dam is converted into electricity. Therefore, the remaining 15% becomes heat. On the other hand, the fraction of electricity transformed into light by incandescent light bulbs is only about 2%, so 98% ends up becoming heat.

The Value of Energy as Heat

Is it impossible for us to use energy once it has turned into heat? In a thermal power plant, fuel is changed into heat, and that heat is transformed into electricity. So clearly heat can be and is used as a source of energy. However, there is probably no easy way to use the heat energy in the air warmed by an incandescent bulb to just a little higher than the ambient temperature. In other words, there is heat that can be used and heat that cannot.

The value of energy as heat is rather difficult to understand, and for a long time scientists puzzled over it. The conclusion finally reached forms one of the basic principles of thermodynamics. That principle is: "heat with a sufficiently high temperature has value comparable to work, electricity and other forms of energy, but as the temperature of the heat gets lower, the value decreases, and heat at the same temperature as the surrounding environment cannot be used at all and therefore has no value." To boil water at 100°C, we want the stove to be at a temperature of at least 150°C, and to melt glass with a melting point of 500°C, we need a furnace with a temperature of 600°C or more. In these cases, the higher the temperature, the better.

Strictly speaking, the value of heat can be described as follows: the fraction of work that can be obtained from an amount heat at a given temperature T is the difference between T and the ambient temperature of the environment divided by T. In other words, the value of heat is the amount of heat multiplied by $(T - T_0)/T$, where T_0 is the ambient temperature. All of these temperatures must be expressed in absolute units. The most commonly used absolute temperature scale is the Kelvin scale. To convert a temperature in degrees Celsius to Kelvin, we just add 273.

For heat at a temperature that is the same as the surroundings, T is equal to T_0, so the value is zero. This means for example that it is not possible to generate electricity using sea water and air at the same temperature. The higher the temperature of the heat, the greater its value, and if the temperature is infinitely high, the ratio becomes one. For example, the sun – one of the hottest things we can imagine – has a surface temperature of about 6000°C or 6273 Kelvin. Using the equation above, we can calculate that more than 95% of heat at the temperature of the sun's surface used in room temperature surroundings could be converted to work.

Let's summarize the main points above. Energy resources from nature are transformed into electricity, gasoline, kerosene, and so on, and those forms of energy are consumed through human activities of "daily life" and "making things." Although saying that energy is consumed appears to contradict the principle of energy conservation, what we mean is that every time energy is transformed, some part becomes heat. And as the temperature of the heat is gradually reduced, its value decreases until finally it reaches the ambient temperature of the environment and loses all its value. You saw in question 3 how heat that becomes the temperature of the environment is radiated to outer space. Therefore, the real nature of energy resource consumption by humans is that through human activities, the chemical energy contained in energy resources such as fossil fuels is transformed, perhaps many times, and each time it is transformed, some part of the energy becomes ambient heat, which is eventually radiated to outer space.

Thinking about energy in this way, we see that the important thing about energy use is not that the quantity of energy is conserved, but rather that energy deteriorates until it can no longer be used. Even though the increase in the amount of energy in the air around an incandescent light bulb is basically the same as the amount of energy in the electricity used by the light bulb, the electricity can be used for many different purposes besides lighting a room, but the energy in the form of slightly heated air cannot be used for anything. This is why humanity is in constant need of new energy resources. However, as we saw in Chapter 1, the fossil fuels upon which we are dependent for almost 80% of our current total energy resources are becoming depleted. Furthermore, the burning of fossil fuels releases CO_2 which brings about global warming.

One solution might be to shift our dependency on fossil fuels to renewable energy or nuclear energy. However, there are also problems associated with using those energy resources. Developing alternative energy resources is certainly important, but completely replacing fossil fuels with renewable energy by the middle of the 21st century is probably not technologically possible, not to mention economically possible. On the other hand, from a safety point of view, it would be best to keep our dependence on nuclear power at a minimum.

So what are the possible roads left to us? This book will suggest the following mid-term and long-term goals. For the mid-term, the goals are 1) to chart a plan for extending the lifetime of fossil fuel resources by limiting the amount of energy used through improved efficiency and 2) to lay out the foundations for constructing renewable energy systems. Once we have achieved these mid-term goals, we must aim for a complete conversion to renewable energy in the long term. In the next section, we will see in concrete terms what it means to improve energy efficiency.

4 Improving Efficiency

Burning Oil Fields Versus Heating Houses

Imagine that an oil field in a desert catches fire and the oil is burned up. The oil turns into CO_2 and water, and at the same time an intense heat is generated. That heat initially raises the temperature of the surrounding air, but in the end the heat spreads out until it is no longer perceptible. Oil turns into heat, and the heat warms the ambient air just the tiniest amount. Thus energy is conserved, but that energy cannot be used to heat a building or drive a car. From the viewpoint of human activities, the energy of that oil has been completely wasted.

Now, consider what happens if we try to warm ourselves using an oil-fired stove in an open field on a winter's night. The oil is burned, becomes heat and warms the surrounding air just a bit, which is the same as what happens in a burning oil field. However, to the extent that we can warm ourselves with the heat from the stove, we derive some benefit from the chemical energy of the oil that is consumed. Of course, if possible, we should put up a tent or some other structure to make it more difficult for the heat to escape, thereby reducing the amount of oil we must burn to stay warm.

When we heat our home with an oil stove, to the extent that we are just burning oil, the situation is the same as a burning oil field or an oil-stove in an open field. However, by burning the oil in a stove in a well-insulated home, we can achieve the goal of warming ourselves with much less oil. This is the essential point of using energy efficiently: we should use the minimum possible amount of an energy resource to achieve a certain goal.

A Vast Range of Efficiencies

Based on the ideas above, let's consider what kind of room heating system has the highest energy efficiency.

We can heat a room using an electric heater, and in that case the heater will produce heat in the same amount as the electricity consumed. So which is more efficient – an electric heater or an oil stove? To answer this, we must determine which option consumes the least energy resources. The oil stove consumes oil to produce heat, and the electric heater consumes electricity. But as we saw earlier, electricity is not an energy resource. To produce the electricity used in the electric heater, fossil fuels must be burned at the power plant. Therefore, we must compare the amount of oil consumed at the power plant to produce the electricity used by the electric heater with the amount of oil burned in the oil stove. Even state-of-the-art oil-fired power plants convert only about 40% of the chemical energy in oil into electricity, and that electricity must then be delivered to your home, which results in an additional loss. Therefore, an electric heater has only 40% of the efficiency of an oil stove.

Recently, air conditioning units that can heat as well as cool a room with electricity are becoming widespread. You might have thought that there is an electric heater in the air conditioning unit, but that is not the case. We will look at the mechanism in detail in Chapter 5, but basically a motor is used to transform the electricity into work, and the work is used to pump heat up to the room from the outside. To "pump up" heat means that even though the outdoor temperature is lower than inside, heat can be moved from outdoors to indoors. Because this is similar to the way that water is pumped up from a low place to a high place, this system is called a "heat pump." In summer, an air conditioning unit uses a heat pump to move heat from the cool indoors to the hot outdoors. In winter, the direction of the heat pump is reversed, so a single air conditioning unit can be used for both heating and cooling.

A heat pump can transport an amount of heat from a low-temperature place to a high-temperature place, an amount of heat several times more than the amount of electricity consumed. Among newer high-efficiency air conditioning units, there are models for home use that can supply an amount of heat to a room that is more than seven times greater than the amount of electricity consumed. The efficiency of transforming oil into electricity is 40%, so an electric heat pump can supply almost three times the heat of an oil-burning stove using the same amount of fossil fuel. The capacity to supply three times as much heat using the same amount of an energy resource may seem like magic, but it is just basic thermodynamics. And more and more of us are doing this when we purchase combination heating and cooling air conditioning units.

If we compare the efficiency of heaters from the point of view of fossil fuel consumption, electric heaters have the lowest efficiency, heat pumps in air conditioning units have the highest efficiency, and oil-stoves are in between. In the case of electric heaters, at even state-of-the-art oil-fired power plants already 60% of the chemical energy of the oil is lost as heat, so only the remaining 40% can be used to heat the room through direct conversion of electricity to heat. Compared to this, an oil-stove that transforms the fossil fuel resources directly into heat is the better choice. Alternatively, since all we are doing is converting electricity into heat, if we consume the same amount of electricity operating televisions, radios, lights and other appliances, we saw in question 1 that we will get almost the same heating effect. And this option gives us more benefits from energy than just running an electric heater.

A heat pump also turns the electricity consumed into heat, but at the same time it pumps several times more heat from outdoors to indoors, so the efficiency is even higher. This example illustrates how, through improved technology, we can reduce the energy resources required to achieve a given goal. With just a tiny fraction of the oil consumed when we try to warm ourselves with an oil stove in a snowy winter field, we can operate a heating system that could comfortably warm a room using a heat pump.

In fact, we can use technology to reduce the energy used for heating and cooling even more. One way is to improve the insulation of our homes by using high-performance insulation in the walls, floors, and roofs and by installing double-paned windows. By improving the insulation of our homes, we reduce the energy demand for heating and cooling. In the Rocky Mountains, at an altitude of 1500 meters, a well-insulated house was constructed in which people lived without consuming any fossil fuel resources for heating and cooling.

Other ways to reduce energy used for heating and cooling include innovative placement of windows under long, sloping roofs, such that in winter when the sun is low in the horizon, sunlight shines into the home and heats it, but in summer, when the sun is higher, the rooms are shaded by the overhanging roof. Planting deciduous trees on the south side of a home is another way to save energy because in summer the leaves block the sunlight while in winter, when the leaves have fallen, sunlight shines into the home. Using fans to circulate air in a building can reduce the cooling load tremendously in summer. And in fact by designing a building in the right way, a natural circulation can be induced so that it is not even necessary to use fans.

Even though the example of "burning oil fields" is rather extreme, in the sense that a fossil fuel resource is burned and ends up only heating the environment an imperceptible amount, there is no difference in principle between burning oil in an open field and heating a well-insulated home. However, the amount of energy resources consumed to gain the same amount of benefit is dramatically different depending on the method used. From the discussion above, we can see that there are three methods for increasing energy efficiency to reduce the amount of resources consumed. First, by using devices such as the heat pump, we can reduce the amount of energy resources consumed to provide some service such as heating, resources that are eventually lost as ambient heat. Second, we can try to do as many things as possible with the energy before it becomes ambient heat, such as turning on televisions and lights to heat a room. Third, we can reduce the amount of energy required to fulfill our needs, for example by insulating our homes and designing homes to get optimal use of sunlight to reduce the energy needed to heat our home. By such methods, we can chart out a plan for increasing energy efficiency to save fossil fuel resources through technology.

From the point of view of efficiency, there is plenty of room for improving the ways in which energy is used. In Chapters 3 and 4, we will see just how large the potential for conservation of energy by increasing efficiency is.

Chapter 3
The Limit of Energy Efficiency

1 The Elementary Steps of Human Activities

How much conservation of energy is theoretically possible? To answer this question, we must first know the minimum energy required to perform a particular activity. The difference between this minimum and the current amount of energy consumed for the activity would be – in theory – the maximum amount of conservation possible. Applying principles of mechanics and thermodynamics, we can obtain this theoretical value for energy conservation. One way to do this would be to calculate the theoretical minimum energy for each energy-consuming human activity, such as the production of steel, the manufacture of plastics, and the use of air conditioners, refrigerators and automobiles. This approach, however, would require studying a countless number of activities. Therefore, let's take another approach here. We will break down the complex human activities into elementary steps and then study the activities where we wish to conserve energy as a sequence of those elementary steps.

As an example, let's consider the process of manufacturing plastic products from oil. The process is comprised of the following parts.

- Oil that is pumped from the oil fields is transported by pipeline to the harbor, loaded into a tanker, and shipped to the region where the plastic is manufactured, where it is transported again by pipeline to a refinery.
- At the refinery, crude oil is separated into various component materials such as gasoline, kerosene, and heavy oil. One of these components, naphtha, is the raw material for plastic.
- Naphtha is heated in a combustion furnace, where through a chemical reaction called pyrolysis, or thermal cracking, compounds such as ethylene and propylene are formed.
- The product of thermal cracking is cooled to around −100°C, compressed and liquefied, and separated by distillation into various component compounds.
- These various components are then further processed into various kinds of plastics and synthetic fibers. For example, ethylene, one of the component compounds, is placed under high pressure and converted through the chemical

reaction of polymerization to grains of a macromolecule called polyethylene. Those grains of polyethylene are then melted and molded to create the polyethylene plastic products and containers you see in stores, such as shampoo bottles and children's toys.

Looking at the description of the process of manufacturing plastics in the previous paragraph, we see that we can break down this process into the following elementary steps: transportation, separation, combustion, heating and cooling, compressing, liquefying, melting, chemical reactions, and shaping. In fact, if we look at the various human activities of "making things" and "daily life" from the viewpoint of energy, almost all of them can be broken down into a combination of some of the elementary steps in the sequence above. We can even break down the human activity of making drip coffee this way. Making drip coffee proceeds through the following steps: coffee beans are transported from some location such as Brazil, roasted, ground up, and finally water is heated and percolated through the grind to make coffee. Therefore, making coffee can be broken down into transport, heating, shaping, heating, and separation.

If we can determine the theoretical minimum amount of energy used in each of these elementary steps, we can easily find the theoretical minimum energy consumption for any kind of human activity by considering it as a combination of the elementary steps. Next, we will estimate the theoretical minimum energy for each elementary step.

2 The Energy of Elementary Steps

The Energy of Transportation Is Zero

First let's consider how much energy is required in the ideal case to transport materials, products, people and so on. As our first example, imagine a car traveling on a level road. To start the car moving, energy is needed. This is because the law of energy conservation states that in order to give objects kinetic energy, the energy of motion, work is necessary.

However, after starting the car and reaching a constant travel speed, theoretically we do not require any more energy to keep it moving. Think back to the speed skating event at the Nagano 1998 XVIII Olympic Winter Games. The gold metal winner, Hiroyasu Shimizu, after reaching the goal, took off his goggles, took off his hat, waited anxiously for the record to appear in the display panel, checked his score, thrust out his fist in exhilaration, and finally stopped moving when he was hugged by his coach. During the whole time he was moving, he did not kick his foot once. Then in the Turin 2006 XX Olympic Winter Games, Shizuka Arakawa performed her signature "Ina Bauer" to win the gold in figure skating. Both of these movements were possible because the friction of ice is small. If there were no friction at all, it would be possible to circle a skating rink that is properly banked for all of eternity without slowing down. Telecom satellites and the moon orbit the earth without stopping, and the earth has continued to orbit around the sun since its formation because there is essentially no friction in outer space.

So what happens when the car stops? If we use a brake to stop the car, the kinetic energy of the moving car turns into heat and ends up warming the air just a little bit. We saw in the last chapter that kinetic energy has the same value as fuel energy, but stopping a moving car in this way causes all of its value to be lost. This is just like the "burning of oil fields" – energy is just wasted. In order to deplete the car of its kinetic energy in a less wasteful fashion, we could force the car to turn an electric generator and transform the energy into electricity. Remember the bicycle with a generator-type light that we saw in the last chapter? Pedaling the bicycle becomes harder when the generator light is turned on, and if you stop pedaling, the bicycle will quickly come to a stop. Therefore, we see that the generator light can function as a brake.

Instead of using a light, let us suppose that we store the generated electricity in a small rechargeable battery. The amount of this electricity will be the same as the kinetic energy that was lost by the bicycle, which is also the same as the amount of work needed to get the bicycle moving again. Therefore, if we use this electricity to drive a motor, we can accelerate the bicycle back to the same speed at which it was traveling before we stopped it (remember that we are considering the ideal case without any friction, but in reality some kinetic energy is always lost to heat in any transformation). Once the electricity is transformed back into kinetic energy, the bicycle will move at a constant speed without any input of energy, and when we want to stop, we can just use the generator to recapture the electricity. In other words, we can make a bicycle that can be started and stopped without having to pedal. And we can think about a car or a truck in exactly the same way. Therefore, we can see that the theoretical minimum amount of energy required for transport on a level surface is zero.

Next, as an example of vertical transportation, let us consider how much we can reduce the amount of electricity required to move an elevator up and down under ideal conditions. You might think that when an elevator goes up, a wire attached to the elevator is wound up using a motor so that electricity is required, and when an elevator goes down, it falls by its own weight, so no energy is needed. However, modern elevators do not work in such a wasteful manner. In elevators, the wire hauling the elevator car up is attached to a pulley, and the other end of the wire is attached to a block having the same weight as the elevator car. Both sides of the pulley have the same weight, and if the pulley is made using high quality bearings so that friction is nearly zero, no energy is required to move the elevator car up and down. In other words, the minimum energy to raise and lower an elevator is zero.

For the transport of oil and natural gas by pipeline, if the diameter of the pipe is increased, the transport friction will become smaller, and at the theoretical limit, the energy required is zero. Even if the pipe goes up and down mountains and valleys, as long as the starting and ending points are at the same height, no energy is required. Think of using a siphon to draw water out of a bath tub into a bucket on the bathroom floor. As long as the outlet end of the hose is lower than the inlet, no matter how high the hose must go to get over the side of the tub, water will flow out of the tub and into the bucket. Energy loss occurs during the transmission of electricity as heat generated by the resistance of the transmission wire. This is the same as the mechanism that an electric heater uses to generate heat. However, without even bringing up the example of superconducting power transmission, we can see here as well that by making the transmission wire "thicker" and the resistance smaller, less heat will be generated. At the theoretical limit, the energy lost during transmission is zero.

From materials to electricity, the theoretical minimum amount of energy consumption for transportation is zero. The main reason that energy is consumed in transportation today is friction. Therefore, the key to reducing energy consumption by transportation is seeing how far we can reduce friction. This is an important point that we will come across again in the next chapter when we consider ways for making passenger cars more energy efficient.

Energy Is Needed for Separation

We saw earlier that separation is an important elementary step in the manufacturing of plastics. In fact, separation is used in all kinds of manufacturing processes, from separating mineral ores from rock to extraction of food seasonings from fermented liquids. Concentration is one form of separation, an example of which is the production of distilled spirits by concentrating the alcohol from fermented alcohol. Also, laundering is the separation of dirt from clothing. Coffee is made from the separation of the coffee component from coffee beans, and butter is obtained by separation of fat from milk. These examples show us that separation is an important step both in "making things" and in "daily life."

In order to separate a mixture into its components, energy is always required. For example, the minimum energy to separate fresh water from sea water is the product of a pressure of 24 atmospheres and the amount of fresh water produced. Let's use this example to see how much energy is needed for separation.

If we partition sea water and fresh water in a container with a cellophane-like semi-permeable membrane that permits water to pass through but not salt, fresh water will seep into the sea water side due to osmotic pressure, and the level of the sea water side will rise above the fresh water side. Osmotic pressure depends on concentration, and in the case of sea water, it is about 24 atmospheres. This means that if we apply a pressure of 24 atmospheres on the sea water side, fresh water will stop seeping through the membrane. If we apply even more pressure, fresh water will seep through the membrane from the sea water side. This way of producing fresh water is called the reverse osmosis method for desalination of sea water.

The amount of energy consumed to produce some amount of fresh water using the reverse osmosis method is determined by the product of the pressure applied and the volume of water obtained. Therefore, the energy used to produce fresh water is proportional to the pressure applied to the sea water side. The theoretical minimum energy is achieved when the pressure is 24 atmospheres, but if we apply just this pressure, fresh water will not actually be produced. If we apply a little more pressure, fresh water will start to seep through the membrane. In actual applications, a pressure of about 80 atmospheres is applied in order to produce a vigorous flow of fresh water. However, to do this, energy is consumed at a rate of 80 divided by 24 or 3.3 times more than the theoretical minimum. The same amount of water is produced, so where did the extra energy consumed go? As in the examples that we have seen before, it is turned into heat and ends up radiated to outer space.

For most kinds of separation in "making things" and "daily life," as much as ten to twenty times more energy than the theoretical minimum is consumed in actual processes. And in all of these cases, the common result of attaining a sufficient rate of separation is the generation of waste heat. Many researchers are working hard to find ways to reduce the amount of excess energy required to attain sufficient rates of separation. For example, one reason that such a large excess pressure is required for desalination of sea water is that the resistance of the separation membrane is large. Therefore, the development of a strong, thin semi-permeable membrane will help us to approach the theoretical minimum of 24 atmospheres of pressure.

Various methods of separation, such as distillation, adsorption, and ion exchange, are used for a variety of purposes, but the theoretical minimum energy required is the same for all of these methods. In fact, the theoretical minimum value does not even depend much on the kind of material to be separated. The main factor affecting the theoretical minimum energy required for separation is the concentration of the different components to be separated. For example, the energy needed to separate the 3% salt content in sea water is about the same as the energy to separate a 3% mixture of CO_2 in the flue gas of a power plant. However, the energy to separate the three parts per billion of uranium in sea water is orders of magnitude greater.

The Energy of Shaping and Forming Is Zero

Putting grains of plastic into a mold to form the frame of a television and pressing a thin sheet of steel into the proper shape for the body of a car are examples of shaping and forming in manufacturing. The theoretical minimum energy required for all of these processes of shaping and forming is zero. This may be difficult to believe, but think about it in the following way. If we heat a material to close to its melting point, it will get soft and easy to shape. If we then recover the heat when we cool the material back down by using an infinitely long heat exchanger to transfer all of the heat of the material to some liquid material, the amount of heat that is recovered will be the same as the energy required for heating. Even though it is not possible to convert all of the energy of heat into electricity or work, in the ideal case it is possible to transfer all of the heat from one material to another. By using that recovered heat to heat up the next material and repeat the same process, we do not need to use any energy. Likewise, the theoretical minimum energy for other forms of shaping and forming, such as making thick plates of steel into thin sheets, cutting and sectioning, and so on, is zero.

Heating and Cooling Using an Ideal Air Conditioner

You might think that if we boil water using a gas flame, as long as the heat of the flame is completely transmitted to the water, in other words, as long as there is no heat loss, we will achieve the highest energy efficiency possible. However, remember that the chemical energy of fuel gas that can be transformed into electricity or work is considerably more valuable than heat energy in the form of water boiling at 100°C or a bath heated to 40°C. Therefore, using fuel to boil water is a huge waste of valuable chemical energy. We saw the same thing when we looked at the different ways of heating a room. The theoretical minimum energy needed for heating and cooling can be determined by considering an idealized form of the common-place air conditioner that we use to cool (and sometimes heat) our homes.

The theoretical minimum amount of energy that is required for cooling was first made clear through the principles of the reverse Carnot cycle in thermodynamics. According to those principles, the minimum amount of electricity needed to pump out a certain amount of heat is determined just by the temperature inside and outside the space to be cooled. The equation that gives this minimum amount of electricity is the temperature difference between the warmer and the cooler spaces divided by the temperature of the cooler side. This is almost the same as the equation that gave us the value of heat in the last chapter, but in this case the denominator is the cooler temperature. Like in the previous equation, all of the temperatures must be expressed in the absolute temperature scale or units of Kelvin, which means we must add 273 to the temperature in Celsius. If the room temperature is 28°C and the outside temperature is 35°C, the value given by this equation is 7/(28 + 273) or 1/43. Therefore, we only need to supply an amount of electricity equal to one forty-third the amount of heat to be pumped out. This is the theoretical minimum for cooling at this temperature.

When we cool a room with an air conditioner, hot air is produced at the outdoor unit. From the point of view of the outside air, this is a heating effect. In other words, we can think of an air conditioner as consuming electricity to take away heat from the air in the room and use it to warm the outside air. The theoretical minimum amount of electricity that must be consumed to heat the outside air a certain amount is also determined by the inside and outside temperatures through the ratio of the temperature difference and the temperature of the hotter side. Therefore, an amount of electricity equal to 7/(35 + 273) or one forty-fourth the required heat is sufficient theoretically to heat the outside air.

Do Compression and Expansion Slowly

It is easy to see that energy is needed in order to compress air. However, the amount of energy depends on the way that the air is compressed. For example, imagine compressing air inside a syringe by covering the tip with your finger. If you press the plunger slowly, the repelling force will gradually get stronger. The energy needed to press down the plunger in this way is close to the minimum. If you press the plunger quickly, from the start, you will feel a strong repelling force, and consequentially the energy consumption will be larger.

The theoretical minimum energy does not depend much on the kind of gas to be compressed, but rather on the ratio of the pressure before and after the compression. Furthermore, the theoretical minimum energy required for compression is exactly equal to the maximum energy that can be obtained during expansion. This is another example of the law of energy conservation.

In summary, we see that the theoretical minimum energy for heating and cooling is determined by the temperature difference, for separation by the concentration of the components, and for compression and expansion by the pressure ratio.

Measuring Chemical Reactions Through an Ideal Electric Cell

We saw in Chapter 2 that all actions can be divided into actions that occur naturally or spontaneously and actions that do not occur naturally but rather require energy to proceed. A stone falls spontaneously if we drop it, but it will not rise unless we provide energy to lift it. Furthermore, we saw that while energy is required to make non-spontaneous processes occur, spontaneous processes can be used to generate energy. Chemical reactions can also be divided into spontaneous reactions such as polymerization and non-spontaneous reactions like the pyrolysis of naphtha. Like all spontaneous processes, spontaneous reactions can produce useful energy such as work or electricity when they occur, and like all non-spontaneous processes, non-spontaneous reactions require energy to occur.

The theoretical value corresponding to the maximum efficiency for chemical reactions depends on the kind of reaction. In spontaneously occurring reactions, those that produce energy, maximum efficiency means getting the maximum amount of energy from the reaction. In non-spontaneous reactions, those that require energy, maximum efficiency means using the minimum amount of energy needed to drive the reaction. Combustion is one kind of spontaneous chemical reaction. By including the reverse non-spontaneous reaction, called reduction, we can discuss the efficiency of combustion as a chemical reaction.

The electrolysis of water to produce hydrogen that we saw in the previous chapter is an example of a non-spontaneous reaction that does not proceed without the addition of energy. The electrical energy that is used during electrolysis can be calculated by multiplying the voltage, the current, and the time. The product of the current and the time of the electrolysis is the amount of electrons used, which determines the amount of water that is split. Therefore, the electrical energy that must be consumed to split a certain amount of water through electrolysis is determined entirely by the voltage, just the same as heating and cooling are determined by temperature, separation is determined by concentration, and compression is determined by pressure.

There is a certain minimum voltage that must be applied for the electrolysis of a particular chemical compound to occur. At any lower voltage, electrolysis does not occur. For water, this voltage is 1.23 volts. The electrical energy consumed at this voltage is then 1.23 volts times the amount of electrons used, and because the voltage is the lowest possible value, this is the theoretical minimum energy consumption for electrolysis of water. However, at this voltage, hydrogen is not actually produced. In order to get hydrogen to form, a little more voltage must be applied. Just as we needed to increase the pressure for the desalination of water, to obtain a sufficient rate of hydrogen production, we need to apply a voltage of about 1.5 volts. However, if we carry out the electrolysis process at 1.5 volts, an amount of electricity equal to $(1.5 - 1.23) \times$ (amount of electrons) is wasted. As before, this electricity turns into heat through the "friction" in the process and ends up disappearing into outer space.

As we can see from our discussion of fuel cells in Chapter 2, a fuel cell works in the reverse of the electrolysis of water. Therefore, once we pool up some hydrogen and oxygen by electrolysis, if we just connect a light bulb in place of the electric power source for the electrolysis, the mechanism of the apparatus will be changed such that the hydrogen and oxygen will be consumed, and electricity will be produced to light up the light bulb. Electrolysis is a process that changes water into hydrogen and oxygen against the natural flow, a non-spontaneous reaction, so energy is required. However, the reaction of hydrogen and oxygen in a fuel cell proceeds without input of energy and can be used to generate electricity, so the fuel cell reaction is a spontaneous reaction. Furthermore, the theoretical maximum energy that can be generated from some amount of hydrogen and oxygen by the spontaneous reaction in the fuel cell is equivalent to the theoretical minimum energy for the non-spontaneous reaction of electrolysis required to produce the same amount of hydrogen and oxygen.

We can obtain the theoretical maximum energy efficiency for any chemical reaction in the same way as for hydrogen and oxygen in the previous paragraph. The amount of energy that must be applied to the form of the reaction that goes against the natural flow (which is the same as the maximum amount of energy that can be extracted from the form of the reaction that goes with the natural flow) can be calculated from the voltage of an ideal electric cell using that reaction. For example, the theoretical minimum energy to make iron from iron oxide is equivalent to the energy to electrolyze the iron oxide with the minimum required voltage. Similarly, the theoretical maximum energy that can be obtained from the combustion of methane is equivalent to the amount of electrical energy can be generated at the maximum voltage of a fuel cell that uses methane in place of hydrogen.

The Theoretical Efficiencies of Energy Devices Are All the Same

A point to stress here is that the theoretical maximum efficiency of these different processes does not depend on the actual method used. For example, once we decide to use methane to produce energy, whether we do so using a fuel cell, a thermal power plant, or a methane engine, the maximum efficiency is the same. Electricity and work have the same value because theoretically one can be converted into the other 100%, so the theoretical maximum amount of electricity that can be produced by a fuel cell or a thermal power plant and the work that can be done using an engine are the same. In concrete terms, the amount is equal to the chemical energy of the methane. In other words, theoretically there is no difference in efficiency between generation of electricity by a fuel cell and by a thermal power plant. So the important question is which technology can come close to this theoretical ideal value the most easily?

The combustion of methane is an example of energy production, but we can think in the same way about the case where energy is consumed. We have seen how we can desalinate sea water using reverse osmosis, but we can also desalinate sea water by evaporating it and then condensing the fresh water. If we carry out this method ideally, the energy required will be exactly the same as using 24 atmospheres of pressure in reverse osmosis. Of course, if we were to simply burn oil and use the heat to evaporate the sea water, and then cool the water vapor until it condensed into water, this would be like warming ourselves with an oil stove in an open field. A thorough effort to make the process consume as little energy as possible is a necessary precondition for approaching its theoretical maximum efficiency.

To summarize, whether we generate energy or use it, if we carry out the process ideally, the amount of energy will be the same whatever mechanism we use. Theoretically, the efficiency of a process involving the transformation of energy does not depend on the actual mechanism of the energy transformation.

Comparing the Energy Consumption of the Elementary Steps

When we burn carbon with the oxygen in air, energy is produced, and in order to remove the oxygen from the CO_2 that is created, energy is needed. These energies are called the energy of combustion and the energy of reduction, respectively, and as explained above, they are theoretically the same. So which is larger,

Table 3-1: The size of theoretical values for energy inputs and outputs in units of kJ/mol

	chemical energy	rxn	evaporation	compression	melt	heating/ cooling	separation	transport/ shaping
Ethyl Alcohol	1278	69	38.6		5.0	2.1–10.0	0.13–1.7	0
Ethylene	1324	136	13.5	5.7–11	3.4	0.9–3.9	0.13–1.7	0
Benzene	3267	208	31.7	5.7–11	9.8	2.5–12.0	0.13–1.7	0
Hydrogen	242	84	0.9	5.7–11	0.1	0.4–1.9	0.13–1.7	0
Iron	412		354		15.1	0.5–22.0	0.13–1.7	0
Aluminum	838		291		10.7	0.5–22.0	0.13–1.7	0

Note: "rxn" is reaction energy and "melt" is melting energy. Reactions are dewatering of ethyl alcohol, hydrogenation of ethylene, hydrogenation of benzene and reduction of copper oxide with hydrogen; compressions are for pressure ratios of 10 and 100; heating is from 25°C to 100°C using 100°C heat; cooling is from 25°C to −100°C using −100°C coolant; separation is for mixtures of 1%/99% and 50%/50%.

the combustion energy of a material or the energy of separation that is required to remove impurities contained in the material? You might think that this kind of comparison is impossible to generalize, but in fact the combustion energy is almost always larger.

Table 3-1 summarizes the theoretical minimums for the amounts of energy consumption required for the elementary processes of several different materials. Based on a consideration of these examples, we will be able to establish rough measures of the size of energy required for each of the elementary processes.

The chemical energy contained in a certain amount of ethanol called a "mole," which is about 60 ml, is 1278 kJ. The energy to separate a mixture of 1% water in a mole of ethanol is 0.13 kJ, which is just one ten-thousandth of the chemical energy. For a 50% mixture of water and ethanol, the separation energy is 1.7 kJ, or 1/705. For normal concentrations of impurities like these, the energy of separation is generally hundreds to thousands of times smaller than the chemical energy.

Furthermore, if we look at the ratio between the chemical energy and the heat of melting for ethanol, aluminum and iron, the values are 256, 78, and 27, respectively. Therefore, the chemical energy is several dozen to several hundred times larger than the energy needed to melt even metals such as iron.

Providing a rough measure of the size of energy going in or out of a particular process is helpful when considering complex energy problems. Of course the chemical energy depends on the molecular composition, and the reaction heat depends on the kind of reaction. The heats of vaporization and melting change according to the type of material. The energies of separation and compression do not depend much on the kind of material, but they are conditional on the concentrations and pressure ratios. However, we can still provide a clear measure of the approximate amounts of energy for each elementary process. By assigning a scale of 1000 to the chemical energy of a material, we can estimate that the approximate order of the theoretical energy consumption is 1000 for combustion and reduction, 100 for other chemical reactions, 10 for evaporation, condensation, compression and expansion, 1 for melting, solidification, heating, cooling, and separation, and 0 for transportation and shaping.

3 The Energy of Human Activities

In the previous section, we have determined the approximate size of the minimum amount of energy theoretically required for each of the elementary steps of human activities. Now let's use these measures to study the activity of making plastic that we looked at in the beginning of the chapter. By doing this, we can determine the theoretical minimum energy consumption needed for the manufacture of plastic by considering it as a combination of the elementary steps above.

First, the energy for transport from the oil field to the refinery is zero. Currently, oil extraction in the Middle East is conducted using a method whereby sea water is injected as oil is pumped up. This is the same principle as attaching a weight to the other side of an elevator car and moving it up and down, so the theoretical minimum energy is zero. The energy for transport by pipeline and tanker is also zero.

Energy is consumed at the refinery during the separation of the crude oil and the pyrolysis reaction of naphtha. The mixture produced by the reaction is compressed and condensed, ethylene is separated, and finally the ethylene is compressed in preparation for the polymerization reaction – all of these steps require energy. During the polymerization of ethylene, reaction energy can be obtained. Finally, the energy for forming the grains of polyethylene that are produced into various products is zero.

We can break down the process of manufacturing plastic into separation, reaction, compression, condensation, separation, reaction and shaping. The approximate measures for these elementary steps are 1, 100, 10, 10, 1, 100, and 0 respectively. Therefore, we see that the largest inputs and outputs of energy are both 100 for the pyrolysis reaction of naphtha and the polymerization reaction of ethylene. When we do the actual calculations, we find that the heat of reaction for polymerization and pyrolysis are almost the same and end up canceling each other out. Therefore, the process of making polyethylene from oil does not contain any elementary steps that require a large amount of energy. Theoretically, it should be possible to reduce the additional ton of oil that is consumed in making a ton of plastic to almost zero.

How about the activity of making drip coffee that we saw at the beginning of this chapter? This process consists of the elementary steps of transport, heating, shaping, heating and separation. The size of energy for each of these steps is 0, 1, 0, 1 and 1, respectively. None of the elementary steps that require large amounts of energy such as combustion and chemical reactions are present. Therefore, we can see that theoretically making drip coffee is an activity that should not need to consume much energy at all. When we consider how we burn gas to boil water and use gasoline to transport the beans, it is clear that we are wasting a large amount of energy. We will see why this and other kinds of energy waste happen in the next chapter.

Chapter 4
Energy Conservation in Daily Life

In this chapter, we will take a look at the potential that technology offers for conserving energy use during our "daily life" activities in homes, offices and transportation. Later, in Chapter 7, we will look at these potentials again when we present the basic concepts of Vision 2050. Our proposal for Vision 2050 will take the year 1995 as our baseline year. Therefore, throughout this chapter, we will base our discussion on the state of technology in 1995. Where more recent data is available, we will examine whether we have succeeded in achieving greater efficiency in recent years.

We saw in Chapter 2 that "daily life" activities make up more than half of the total energy consumed by human activities that has been converted into useful forms such as electricity and gasoline by the energy transformation sector. In Chapter 3, we examined the theoretical minimum amount of energy required for these activities. How much energy conservation is actually possible through technology? Let's start by looking at the possibilities for energy conservation in transportation by focusing on the main user of energy, the automobile.

1 The Automobile

In the previous chapter, we saw that, in theory, the amount of energy required for transportation is zero. Thus, ideally it should be possible for passenger cars and other motor vehicles to run without consuming any fuel. But if this is true, how can we explain the fact that consumption of gasoline by motor vehicles today constitutes over 20% of the total energy use by human society? First, we will look at the mechanism by which passenger cars consume gasoline. Once we identify where the important energy losses are, we can decide which methods are most effective in achieving energy conservation in automobiles by determining the methods that can most effectively reduce these energy losses.

Hiroshi Komiyama and Steven Kraines
Vision 2050: Roadmap for a Sustainable Earth.
© Springer 2008

How Conventional Automobiles Work

Passenger cars and other motor vehicles run by combusting fuel in the car engine. When fuel in the cylinders of the engine combusts, the resulting force is applied to the cylinder head, causing the axle to turn. Then, through a multitude of gears and other transmission parts that make adjustments for the speed and direction, the wheels are turned. In this chain of events, the chemical energy of gasoline is changed into work by the cylinder head, and that work is used to move the car.

The first step is the transformation of the chemical energy of gasoline into work and heat. The law of conservation of energy holds, so the sum of heat and work produced by the combustion of the fuel must be equal to the chemical energy of the fuel. Ideally, all of the gasoline should be transformed into work, but in passenger cars today the amount of energy that becomes work is only about 35%. The remaining 65% is lost as heat in the exhaust emissions and radiation from the engine.

To start a car moving, the driver presses the accelerator down firmly—putting the "pedal to the metal." This causes a large amount of gasoline to be combusted in the engine, producing a correspondingly large amount of work. As a result the car obtains kinetic energy, which causes it to accelerate. However, not all of the work generated in the engine is transformed into kinetic energy. Because of various forms of friction, such as the friction between the tires and the ground or the friction between the gears and the transmission, a considerable amount of the work ends up becoming – you guessed it – heat.

Once a car reaches the desired speed, the driver does not need to press the accelerator down so far because, in comparison to putting the car into motion, keeping it in motion takes less energy. However, we saw in the previous chapter that ideally no energy should be required to keep it moving at a steady velocity. So if we are traveling at a constant speed on a level road, why do we need to consume any gasoline? The culprit is friction. Once again a large part of the car's kinetic energy ends up becoming heat, through friction between the tires and the road and from the gears inside the car. Furthermore, when we are driving at faster speeds, like on a freeway, friction between the car body and the air becomes significant, producing even more heat.

Another problem is that, although the maximum efficiency for a gasoline internal combustion engine is 35%, the actual efficiency of an engine varies according to the driving conditions. Usually, a car engine is designed to have its maximum efficiency under conditions of slightly higher output, such as a moderate acceleration. When less engine power is required, such as during low-speed driving, or when maximum power is produced by pressing the accelerator to the floor, the efficiency decreases.

To stop, at a red light for example, the driver presses down on the brake. Pressing down on the brake causes a brake plate to press against the metal part of the car wheel. This results in friction between the brake and the wheel, which slows the car. As we saw in the previous chapter, the ideal way to slow a car would be to using a generator brake collect the kinetic energy of the moving car as electricity. However, in conventional cars that use brake plates, the kinetic energy of the car ends up transformed into heat through the wasteful process of friction.

What about when we are stopped at an intersection? If the engine is running, then gasoline is still being burned. At this time, all of the work created by running the engine ends up heating the engine oil, the gears and the air, and then disappearing as waste heat.

In summary, there are six factors that together explain why, even though the energy for driving a car should theoretically be zero, such a large amount of energy is consumed in reality.

1) The efficiency in transforming chemical energy to work is not 100%; some chemical energy of the fuel combusted in the engine ends up as heat that disappears into the environment.
2) Friction in the gears and moving parts of the car generate heat during the transmission of work from the engine to the tire.
3) Friction between the tires and the ground generates heat.
4) Friction between the car body and the air generates heat.
5) Friction in the brakes generates heat.
6) An idling engine results in a waste of energy.

Improving automobile technology to address these factors should be the guiding principle for improving energy conservation in cars.

High-Efficiency Engines and Hybrid Cars

One way to raise the efficiency of transforming chemical energy into work is improve the engine. In internal combustion engines, fuel is combusted in the cylinders of the engines, providing force to drive the automobile. To obtain the most force from the combustion of fuel in the engine cylinders, the gasoline needs to be vaporized and mixed with air. In gasoline engines made in the 1990's, gasoline taken into the cylinder was vaporized using the principle of "atomization." Atomization is the same process used to vaporize perfume in a perfume spray bottle. When liquid mixed with air is forced through a small opening, the liquid turns into gaseous form. Gasoline was atomized in the car engine by forcing it through a valve called a carburetor. The mixture of air and fuel was forced through the carburetor using work from the expanding cylinder; therefore that amount of work had to be subtracted from the work generated during combustion to get the net output of the engine. At low driving speeds, the amount of gasoline consumed was decreased by partially closing the carburetor, which increased the amount of work required to force the air and gasoline through. As a result, the decrease in engine efficiency was especially large at low driving speeds for engines using carburetors.

To overcome this problem, a new kind of engine, which compresses gasoline and injects it directly into the cylinder, has been developed and marketed in passenger cars sold today. This engine is called a Gasoline Direct injection engine, and it works in the same way as conventional diesel engines. In direct injection engines, fuel is pressurized and then injected into the cylinder. Therefore, at low speeds all we need to do is reduce the amount of fuel that is injected, so no additional work is required to supply the fuel into the engine. With this design, an improvement in efficiency of about 25% has been demonstrated over ordinary gasoline engines. In fact, as of the writing of this book, no more cars are being manufactured with carburetors.

The efficiency of internal combustion engines, both gasoline and diesel, also depends strongly on how much the gas mixture of fuel and air is compressed before it is ignited. The greater the compression, the larger the force of the explosion, and the higher the efficiency. Direct injection engines contribute to increased efficiency in this regard as well, because only the air is compressed by the engine and the fuel is just injected into the compressed air. Air is more compressible than fuel, so the compressibility ratio of the fuel/air mixture can be made higher. Furthermore, through the use of computers to precisely control the injection of air and fuel to the cylinder, it is possible to achieve ultra lean mixtures of air and fuel. Ultra lean mixtures are mixtures of fuel and air where the ratio of air to fuel is considerably more than the stoichiometric combustion ratio, as much as three times more. With so-called "lean burn" engines, even higher compression ratios are possible, making it possible to further reduce the loss of efficiency and pollutant emissions when driving at low power output levels. These are examples of the improvements in automobile technology that have occurred just in the past decade.

The maximum efficiency of diesel engines is 40–45%, which beats the 35% of gasoline engines. However, diesel engines cause environmental problems because the exhaust emissions often contain high levels of soot and nitrogen oxides. To take advantage of the higher efficiency of diesel engines, we must overcome this pollution problem. Some of the new technologies being developed to make diesel engines cleaner include the use of Common Rail Injection to increase the injection pressure of the fuel thereby producing a finer atomization of the fuel, and the improvement of catalytic converters with Diesel Particulate Reduction systems to reduce soot emission.

Although these methods for improving the engine itself are important, there is even more potential for reducing energy consumption by running the engine under the conditions that give the best possible fuel efficiency. The average fuel efficiency under the standard driving conditions in Japan, called the "10–15 mode," is around 13%, which is only about a third of the maximum efficiency of 35%. The reason for the decrease in fuel efficiency is that for much of the time that the car is driven in city traffic, the engine is required to provide power that is either above or below the optimal output level. If we could keep the engine producing power at the maximum efficiency, we could increase the overall efficiency almost three-fold. We could do this, for example, by storing the excess work that is produced during low driving speeds and using it to provide the additional work required for acceleration

Fig. 4-1: The Prius Hybrid Car (Courtesy of Toyota Motor Corporation)

and travel at high speeds. Technologies for improving the engine itself, like the use of direct-injection and lean burn technologies, can increase fuel efficiency by at most 10 to 15%. So an opportunity to triple fuel efficiency is something that we cannot ignore.

Hybrid cars, such as the Toyota Prius and the Honda Insight, attempt to increase fuel efficiency of gasoline engines in this way. A hybrid car is a combination of an electric car and a gasoline car; you can think of it as a normal car with a larger battery and an electric motor. In other words, a hybrid car has two sources of energy for driving: the gasoline engine and the electric motor. When a hybrid car is driven at speeds requiring power output that is close to the optimum output of the gasoline engine, the gasoline engine is used to drive the car. If excess work is produced, the hybrid car uses that work to generate electricity and charge the battery, and if additional work is required, some models of hybrid cars can use the electric motor to supplement the power output of the gasoline engine. When the hybrid car is being driven at low speeds that are not optimal for the gasoline engine, the gasoline engine is turned off and the electric motor powered by the battery is used to move the car. Also, the engine turns off when the car is stopped at a light, and the electric motor is used to start the car moving again. When the car reaches an appropriate drive speed, the gasoline engine is restarted.

Having a larger battery in the car gives us the opportunity to capitalize on another method for conserving energy we have seen, called regenerative braking. Remember the example in the previous chapter of the bicycle that can start and stop without pedaling? "Regenerative braking" means using the electric generator in the hybrid car to convert the kinetic energy of the car into electricity when braking. Therefore, the hybrid car is a design that can contribute significantly to the solution of three of the factors that contribute to the consumption of energy by automobiles: the transformation efficiency from gasoline to work, the friction in the brakes, and wasteful fuel use during engine idling. In locations such as central Tokyo where the driving efficiency of normal automobiles is low due to the traffic congestion, hybrid cars can operate with about half of the amount of gasoline used by conventional cars.

Fuel-Cell-Powered Electric Cars

Many other methods are being studied to increase the efficiency of transforming gasoline into work. As we saw in Chapter 3, the theoretical maximum efficiency for transforming the chemical energy of fuel into work is the same for engines, electrical cells, and thermal power plants – essentially 100%. All we need to do is turn the wheels of an automobile for it to run, and there are many ways to provide energy for doing that.

Proponents of electric vehicles argue that electric vehicles are more fuel efficient than gasoline engine vehicles because the efficiency of electric power plants is greater than the efficiency of gasoline engines. We have seen that the maximum efficiency for conventional gasoline engines is 35%, and that – even with the use of advanced technologies such as direct injection and lean burn – the most that we can hope for in the near future is an efficiency of 40%. Currently, there are thermal power plants in operation with power generation efficiencies of more than 50%. Not only does the generation efficiency of the thermal power plant greatly exceed the maximum efficiency of automobile engines, but because electric motors can be easily started and stopped, electric vehicles also have the advantage of eliminating the loss of energy caused by idling a gasoline or diesel engine when the car is stopped. From the combined effect of these two efficiencies, electric vehicles could contribute considerably to energy conservation in transportation.

Currently, the type of electric vehicle getting the most attention is probably the fuel cell vehicle. There are many types of fuel cells, ranging from ones that operate at temperatures above 1000 °C to ones that run at close to room temperature. One of the fuel cells with the highest potential for being a power source for automobiles in the near future is the polymer electrolyte fuel cell. Polymer electrolyte fuel cells produce electricity from hydrogen fuel at close to room temperature. If hydrogen is loaded on the vehicle and electricity is generated through the reaction with oxygen in air, even now an electricity generation efficiency of 50% is possible.

The development of a commercially viable fuel cell car has yet to be achieved. Many of the problems to be solved are related to the fuel cell itself, such as lifetime, reliability, weight, capacity and cost. However, there are other problems, such as how to set up supply stations for hydrogen fuel. Furthermore, we have to figure out how to store hydrogen on the vehicle. If we store the hydrogen in a tank, the tank would have to be pressurized far higher than a propane tank. Another way is to store the hydrogen within the molecular matrix of a special metal alloy and load that metal onto the vehicle. Alternatively, configurations of fuel cell cars are being studied where methanol, which is a liquid and therefore easier to handle, is loaded onto the vehicle instead of hydrogen. In one configuration, the methanol is transformed into hydrogen for use in the fuel cell. At a large factory, it is possible to make methanol into hydrogen relatively easily and at a high-efficiency. However, in an automobile it is much more difficult. As a consequence, many automobile companies are also conducting research on fuel cells that use methanol rather than hydrogen as the fuel for generating electricity. These fuel cells are called, not surprisingly, direct methanol fuel cells.

These various configurations of fuel cells for cars are currently the subject of intense research and development. It has been estimated that if a fuel cell vehicle with high-efficiency can be developed, it could more than double the current efficiency of transformation from fuel to work.

Lowering Vehicle Weight

The most effective way to decrease friction between the tires of a vehicle and the ground is to make the body of the vehicle lighter. For example, the difference in the effort needed to pedal a high-performance racing bicycle made of light-weight alloys as compared to that needed to pedal a home-use iron clunker is unbelievable. The bicycles used in races such as the Tour de France are truly light-weight – they can be easily lifted with one hand. Using them, the competitors can pedal up and down mountains. With a typical clunker made of iron, even a superhuman competitor could not accomplish this feat. For the same reason, marathon runners are slim and lightweight, not brawny and heavy.

This point is worth emphasizing. Weight reduction is one of the most important keys to reducing the energy consumed in transportation. We can see this in the relationship between the consumption of gasoline and the weight of automobiles, which is almost linear, as shown in figure 4-2. The reason is that friction is proportional to weight. One way to reduce vehicle weight is to reduce vehicle size. However, it is also possible to maintain the size of the automobile while reducing the weight by using special materials such as an iron alloy called "high-tensile

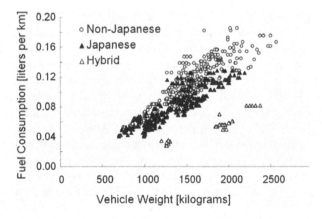

Fig. 4-2: Fuel required for a car to travel 1 km (Data from Yahoo Jidosha)

steel" that has a high strength per unit weight. The weight of vehicles can be reduced even further through the use of lightweight materials such as aluminum and plastics. Automobile manufacturers today are seeking ways to reduce automobile weight without compromising size, safety or performance. It is probably not too much to expect that the weight of passenger cars will be reduced by half in the next decade.

The Future Form of Automobiles

Because in theory automobiles can travel with zero energy, we know that there is a huge potential for reducing energy consumption in car transportation. Let us consider here some possible ways for designing automobiles that can provide the same performance as today's passenger cars while consuming considerably less energy.

First, let's think about the efficiency of a race horse. On September 30th, 2001, Trot Star ran the 1500 meter distance of the Sprinter's Stakes in a record time of 67 seconds. This corresponds to an average speed of 64 km per hour. In other words, a thorough-bred race horse, that is essentially just a single "horse-power," was able to run at a speed that matches the performance of a 100 horse-power car. Another way of saying this is that a horse can run with just 1/100th of the energy of a car. One reason is that horses have far less friction with the ground when they run than cars. A horse obtains propelling force efficiently by kicking the ground with its hooves. In the same way as we saw in the example of the iron ball hitting the ground in Chapter 2, the horse loses some of its kinetic energy as heat to the surroundings when its hooves strike the ground. However, the horse has evolved to run extremely efficiently, so this energy loss is minimal. Because the area of contact with the ground is small and the time that the horse is touching the ground is short, in essence the horse "flies" over the ground.

You might think that if the friction between the tires of a car and the ground were too small, the wheels would spin freely and the car would not move. This could certainly happen in the cars that we drive today. However, a reduction in friction does not necessarily mean a reduction in the propelling force that is transmitted to the ground. One example of a mode of transportation that overcomes this problem is ice skating. In ice skating, you put your weight on the skate on one foot, which allows you to skate with just a minimum amount of friction. You use the skate on your other foot to push against the ice and gain propelling force. Another example is a method for mountain climbing where the fur of a seal, called a "climbing skin," is stretched over regular snow skis. Due to the alignment of the fur, a climbing skin makes it possible to slide forwards but not backwards. In long-distance ski competitions, the same property is achieved through a special way of applying the wax to the skis. If we could develop tires that propel a car in a similar way, we could build a car that travels with greatly reduced friction between the tires and the ground.

Here is another example that shows the importance of the weight of a car. There is a race where cars compete to have the highest fuel mileage. A slender driver operates a car with a light-weight body and thin tires. In 1998, the winning car went 1600 km on a single liter of gasoline. More recently, a fuel-cell-powered car was developed at the Swiss Federated Institute of Technology in Zurich that could go 5134 km using the equivalent of one liter of gasoline. Compared to conventional passenger cars with fuel efficiencies on the order of 10 km per liter, the winners of these fuel efficiency races can operate with $1/160^{th}$ to $1/500^{th}$ the amount of gasoline.

Horses run with $1/100^{th}$ and a fuel-efficiency race winner runs with $1/500^{th}$ the energy of a conventional passenger car. How far can we push energy conservation of cars? By doubling the transformation efficiency from fuel to work and halving the weight of the car, it should be well within the realm of possibility by the middle of the 21st century to manufacture cars that consume only one-fourth the fuel needed in 1995 models. In fact, already hybrid cars get almost twice the fuel efficiency of standard gasoline engine cars, and the introduction of Gasoline Direct injection engines has increased the fuel efficiency of conventional gasoline engines by 25%.

Here is another example. Most automobiles today have an automatic transmission. Automatic transmission engines used to consume about 10% more gasoline than a manual transmission automobile driven by an expert driver. The reason is as follows. In a manual transmission, the clutch connects without any slippage. However, in an automatic transmission, the clutch is always slightly loose, resulting in a small amount of slippage. This slippage causes friction in the car transmission, reducing the fuel efficiency of the car. But with the introduction of continuous variable transmission (CVT) engines, this problem has nearly been solved, resulting in nearly a 10% increase in fuel efficiency.

How about after that? It is probably impossible to create an automobile that runs exactly the same way as a horse. However, the development of tires that can transmit propelling force to the ground with high-efficiency and little friction should certainly be possible. By making many small technological improvements, it might be possible to achieve fuel consumption that is one tenth that of today's automobiles. However, we are unlikely to create a commercially viable passenger car having the 500-fold increase in fuel efficiency of the one-liter race winner. Still, we should not underestimate the potential of technology to make tremendous improvements in efficiency.

This discussion brings to mind the establishment of new sports records. The long believed "human barrier" of 100 meters in 10 seconds was broken in 1968. Following that, the 9.9 second barrier was broken, and in 2005, a record time of 9.77 seconds was set by Jamaica's Asafa Powell. How much further can this time be reduced? Records of 9.6 seconds or even 9.5 sections may be made, but surely no one could run the 100 meter race in 9 seconds flat. Or could they? With the development of a revolutionary training method or the appearance of a sprinter with an order-of-magnitude difference in strength, even the 9 second barrier may be broken. Technology innovation is the same. The possibility for unforeseen discoveries and inventions is ever present. Up until this point, our discussion has been limited to predictable extensions of the current state-of-the-art of technology. However, to the extent that the theoretical energy for transportation is zero, it is impossible to say what the limit of technology is.

2 Homes and Offices

In homes and offices in Japan, energy in the form of electricity, city gas, and kerosene is consumed in nearly equal amounts for three main kinds of "daily life" activities: 1) room heating and cooling, 2) cooking and heating water, and 3) lights and electric appliances. These uses account for half of the energy that is consumed in "daily life" activities, the other half being consumed by transportation. Remembering our discussion of the theoretical minimum energy needed for "daily life" activities, let's look at the difference between the reality and the ideal for room heating and cooling, water heating, and lighting.

How an Air Conditioner Works

As we saw before, a modern air conditioner provides both heating and cooling by using work created from electricity to pump heat up from the lower-temperature side to the higher-temperature side. The mechanism for heating and cooling is the same, so let's use the example of cooling shown in figure 4-3.

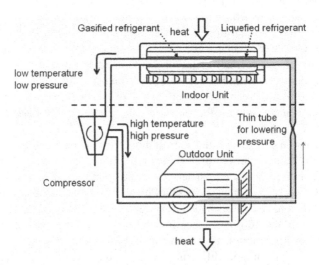

Fig. 4-3: The basic mechanism of a heat pump (example of a cooling system)

Most room air conditioners today are composed of an indoor unit and an outdoor unit. A special liquid called a refrigerant is circulated through a pipe that connects the two units. Wiping alcohol on your skin, for example before getting a flu shot, gives your skin a sudden chill. This cooling effect is caused by the removal of heat from your skin when the alcohol evaporates. In the same way, if the liquid refrigerant evaporates, it will remove heat from the surrounding air. On the other hand, if the gaseous refrigerant is cooled by the surrounding air, its heat will be transferred to the air and it will condense back to its liquid form. We can imagine this by thinking of a window pane in the winter. When it gets cold, lots of dew drops form on the window that can eventually collect to form little streams of water. The reason is that the water vapor in the room loses its heat to the cold window pane, cools, and condenses in the form of a dew drop. An air conditioner running in a cooling mode uses these vaporization and condensation mechanisms to transport heat from indoors to outdoors via the refrigerant.

There is a problem, though, with the mechanism described in the previous paragraph. The refrigerant evaporates at higher temperatures and condenses at lower temperatures, but in that case, heat will be transported from the high-temperature side to the low-temperature side. In the summer, that means we would be transporting heat from the hot outdoors into our home, exactly the opposite of what we want! The way room air conditioners reverse this flow is to make the pressure of the refrigerant in the outdoor unit higher than that in the indoor unit. If the pressure is high, the refrigerant will condense even at a high temperature. The reason a pressure-cooker can cook food more quickly is the same – by increasing the pressure, the boiling temperature of water becomes higher than 100 °C. To increase the pressure, an air conditioner uses a compressor, which consumes electricity. In fact, the electricity consumption of the compressor makes up almost all of the energy consumed by an air conditioner. When liquid refrigerant is returned to the indoor unit, it passes through a thin tube, called an expansion valve, which decreases the pressure. At the lower pressure, the liquid refrigerant evaporates even at the lower temperature in the room. In this way, it is possible to transport heat from the cool indoors to the hot outdoors.

Let's say that you want to use the air conditioner to keep your room at a reasonably cool 28 °C on a summer day with an outdoor temperature of 35 °C. Under these conditions, an air conditioner with ideal energy efficiency would evaporate the liquid refrigerant indoors at a temperature of 28 °C and condense the gaseous refrigerant outdoors at a temperature of 35 °C, using a compressor that requires just the theoretical minimum amount of work to compress the gaseous refrigerant. As we saw in Chapter 3, the relationship between the energy consumption of this ideal air conditioner and the amount of heat pumped out of the room is given by the temperature of the room in absolute temperature units divided by the temperature difference, which is $(273 + 28)/7$, so the amount of heat that can be pumped out of the room is 43 times the amount of work consumed. However, room air conditioners sold in 1995 could pump out an amount of heat from a room that was at most four times the amount of electricity consumed. We have seen that the value of electricity and work is equal, so those models achieved less than a tenth of the ideal efficiency.

Energy Conservation by Improving Air Conditioner Efficiency

There are two main reasons for this gap between the ideal value and the actual value for the efficiency of room air conditioners. The first is that the compressor consumes about twice the electricity theoretically required. This excess electricity is consumed because the efficiency of converting electricity into work and the amount of work used in compression are both much larger than the ideal values. We can improve the efficiency of converting electricity into work by using a high-performance permanent magnet in the motor. By improving technologies, the work used for compression in large-scale compressors, such as those in factories, has already been raised to efficiency levels as high as 90%. The efficiency of compressors in room air conditioners is only 50%, so it should be possible to improve this value. As we saw in Chapter 3, the basic principle for making compression efficient is to do the compression slowly, as reflected in the fact that if you compress the air in a syringe slowly, you can do so with a relatively small amount of work, but if you compress the air quickly, the work required increases greatly. In small-scale compressors, like those used on room air conditioners, there is often no way to avoid doing the compression quickly. However, if we include energy conservation in design goals, we can certainly improve this efficiency.

The second and more important reason that the difference between the ideal and the reality is so large is the size of the temperature difference used in room air conditioners. Although the difference between the indoor temperature of 28 °C and the outdoor temperature of 35 °C in our example is only 7 °C, room air conditioners made in 1995 were designed so that the temperature of the refrigerant was 5 °C in the indoor (cooling) unit and 40 °C in the outdoor (heating) unit – a difference of 35 °C. For this reason alone, more than five times the ideal amount of electricity is required. Combining this five-fold increase from temperature difference with the two-fold increase from compression yields the ten-fold difference we saw before between the ideal efficiency and the actual efficiency of a typical room air conditioner.

The difference between the indoor temperature and the indoor unit is 23 °C. In contrast, the difference between the outdoor temperature and the outdoor unit is just 5 °C. Why is this? Conventional air conditioners improve the heat transfer efficiency in the outdoor unit by using a powerful fan to blow air through the tubes containing the pressurized refrigerant. Because the air flow is so strong, the five degree temperature difference between the 35 °C air and the 40 °C refrigerant is enough for the air conditioner to work. However, there is a downside: when you walk close to an outdoor unit, you are hit by a blast of hot air. If we could make the flow of air in the indoor unit about the same strength as that of the outdoor unit, then a cooling refrigerant temperature of five degrees less than the indoor air temperature would be sufficient. As a result, the temperature difference between the indoor unit and the outdoor unit in our example would be reduced from 35 °C to 17 °C, so we could realize a 50% energy savings.

Recently, manufacturers have been studying ways to improve the transfer of heat in the indoor unit. If we could increase the heat transfer area between the refrigerant and the air, a smaller temperature difference would be enough to supply the required cooling without strengthening the flow of air. Various techniques are used in current air conditioners to increase the heat transfer area. One technique increases the surface area in contact with the air by attaching fins to the outside of the pipe through which the refrigerant flows. Another technique involves attaching baffles on the inside of the pipe, which causes turbulence in the flow of the refrigerant, thereby increasing the heat transfer rate. Also, new configurations such as wall heating and cooling are being tried. If we use the entire area of the wall, a far greater heat transfer area can be obtained, so a sufficient heating and cooling effect can be obtained through a smaller temperature difference. Furthermore, the variation of temperature in the room will be reduced; thus, as a side benefit, we create a more comfortable living environment.

Another way to improve heat transfer in the indoor unit is to design a better flow path through the unit. Today, manufacturers use computer simulation models to plan the best positions for the heat exchanging units inside the air conditioner so as to maximize the transfer of heat.

As a result of these technology improvements, room air conditioners have improved remarkably over the last decade. The newest air conditioners in Japan can pump an amount of heat out of a room that is more than seven times the electricity consumed – an improvement of 40% compared to the highest-efficiency models in 1995. This increase in efficiency has been achieved in part through improvements in the efficiencies of the compressor and other components of the air conditioner. However, even more important were the improvements in air flow that made it possible to reduce the difference between the room temperature and the temperature of the refrigerant in the indoor unit by almost 30%.

However, if we raise the temperature of the indoor refrigerant for cooling in the summer too high, another problem will emerge. Cooling is actually only one of two important services provided by air conditioners. The other is the drying effect. Part of the discomfort that you feel on a hot summer day is from the high tempera-ture, but high humidity also is an important factor. Conventional air conditioners remove not only heat from a room, but also humidity. They can do this because the humidity in the warm air condenses inside the indoor unit and is removed. But this condensation only happens when the indoor unit is sufficiently cold. If the temperature of the indoor unit gets much above 15 °C, the rate of condensation will decrease dramatically.

To answer this problem, manufacturers are designing air conditioning systems that remove heat and humidity in separate stages. Humidity is removed through the use of special materials called desiccants, so the heat pump part of the air condi-tioner only needs to remove the heat from the room. In this way, the temperature difference in the indoor unit can be reduced even more.

Energy Conservation Through Load Reduction

Improving the air conditioner is not the only way that we can reduce the amount of energy consumed in heating and cooling. We can also better insulate our homes. In the summer when you return home, if you have been gone for awhile and the sun has been shining brightly, you may find that your house is stiflingly hot. So you quickly turn on the air conditioner. In just two or three minutes, your house cools down. However, if you then turn off the air conditioner, heat will leak in from outside, and soon it will become hot again. Therefore, it is better to think of air conditioning not as cooling a hot room, but rather as pumping out the heat that leaks in from outside. Insulating your home reduces the amount of heat that leaks in.

The amount of heat that leaks in from the hot outdoors is called the cooling load. The energy needed for cooling is given by the ratio of the cooling load to the amount of heat that can be pumped out of the room using a given amount of energy. Increasing the efficiency of the air conditioner reduces energy consumption by increasing the heat pumped out with a given amount of energy. However, we can also reduce the energy required for cooling by reducing the cooling load.

The amount of heat that flows into a room is proportional to the indoor/outdoor temperature difference; the larger the difference, the greater the flow of heat. So one way we can reduce the cooling load is to raise the thermostat setting; in fact, we can reduce the cooling load to zero by setting the room temperature the same as that outdoors. Energy saving actions such as raising the thermostat in summer are important. However, the focus of this book is on the role technology can play in achieving a sustainable society. Deciding to raise or lower the room temperature to save energy – a problem of lifestyle – is outside the scope of this book.

The technological method for lowering the cooling load is to improve the insulation of the room being heated or cooled. This includes doing things like using high-quality materials to insulate the floors, walls, and roofs, and making windows double-paned. In houses in Northern Europe and Canada, where the winters are severe, many ingenious devices for insulation are employed. However, in the process of insulating our homes, if we end up making them too air-tight, the air inside will get stuffy and stale, so we will need to improve ventilation. Of course, if we just open the windows, the heating and cooling load will increase, defeating the purpose of insulation.

On the other hand, if we allow the outside air flowing into the house to exchange heat with the inside air flowing out through a thin plate of metal, we can use the warm inside air to heat the cold outside air as it flows into the house. Moreover, if instead of metal, we use a separator that allows water vapor as well as heat to pass through and exchange between the incoming and outgoing air, we can dry the air coming in during summer and recover the moisture of the air going out in winter. In fact, almost half of the new office buildings in Japan are equipped with such heat and humidity exchange systems. Residential buildings are also increasingly using such ventilation systems. However, much of the air that is taken out of rooms such as kitchens and bathrooms is not suitable for heat and humidity exchange.

Therefore, the efficiency is not as high in houses and apartments as it is in office buildings.

It Is Wasteful to Use Gas to Boil Water

Next, let's take a look at the consumption of energy for heating water by considering the example of preparing a bath. The process of preparing a bath involves heating 20 °C water to 40 °C, and for the same reasons as in heating a room, the minimum amount of energy required is achieved by an ideal heat pump. If the temperature difference is 20 °C, then at least in theory just 20/(40 + 273) or 1/15th of the amount of heat needs to be supplied as work. In comparison, heating the bath directly by burning some fuel such as gas means that we need to consume at least an amount of fuel energy that is the same as the amount of heat required for the bath. Therefore, we can achieve our goal with far less fuel consumption using a heat pump. This is the same as the reason we noted in the previous chapter that burning gas to boil water in making drip coffee is such a waste of fuel.

Using a heat pump to heat a bath or boil water is more difficult than to heat a room. If we were to make the temperature of the heat pump fluid 45 °C in order to obtain 40 °C water for the bath, it would take too long for the water to heat up. To heat the water fast enough, we must raise the temperature; however, as the temperature difference of the heat pump is increased, and the amount of work required increases. Ten years ago, this might have seemed to be an insurmountable problem. However, through the efforts of electric power companies and manufacturers, heat pumps are now available on the market that heat water from ambient temperature to 90 °C, which is more than enough to supply the hot water needs of homes.

There are other alternatives to reducing the large waste of energy occurring when we heat a bath directly. For example, we know from the law of energy conservation that when we combust energy resources at factories and power plants, even if along the way the energy is transformed into useful forms such as electricity, work and kinetic energy, in the end it all becomes heat. In general, useful forms of energy cannot be obtained from heat that is at a temperature of around 40 °C, so at a factory, there are countless sources of excess heat at these relatively low temperatures. We can think of this low-temperature heat as a waste product of energy resources, and in fact we often speak of waste heat being dumped into rivers by thermal power plants. If we were fortunate enough (or unfortunate enough!) to be living near a factory or power plant, we could use their low-temperature heat to heat our bath water.

How Effective Is Cogeneration?

Another possibility that has been suggested for obtaining heat with less waste is a method called cogeneration. Cogeneration, sometimes called "combined heat and power," means the simultaneous generation of both electricity and heat.

In Chapter 2, we saw that there are two ways to generate electricity – using a generator or using an electric cell. Many different techniques are used to generate electricity in both of these ways, including gas turbines or fuel cells. Each technique loses some amount of the input energy as heat. For example, in the process of generating electricity using a gas turbine, a large amount of the chemical energy of the fuel becomes heat. Cogeneration tries to put both the heat and the electricity to effective use. If the heat from the gas turbine is released at a temperature of 100 °C, that is sufficient for heating bath water, making coffee, and providing hot water for other uses in homes and offices. In this way, we could make use of heat that would otherwise have dissipated into the environment.

In reality, cogeneration has not succeeded as well as expected. The main reason is that, compared to the demand for electricity, the demand for heat is small. In cogeneration systems based on gas turbines at the turn of the century, 30% of the chemical energy of the fuel is made into electricity, 40% into heat, and the remaining 30% is lost. However, there are only a few places where more heat is used than electricity, such as hotels with heated pools. If we are not going to use the heat anyway, then it is better to use standard electric power plants, which have electricity generation efficiencies of more than 50%. Even if we include the heat from cogeneration, a cogeneration system that produces 30% electricity and 40% heat at 90 °C, is a worse deal in terms of resources consumed than an electric power plant that generates 50% electricity and throws away the rest of the heat. The reason is that even if we end up using 20% of the electricity generated by the standard power plant to produce heat, we can use heat pumps available on the market today to pump up three times as much heat from ambient temperature to 90 °C. That is equivalent to 60% of the original chemical energy, which exceeds the 40% heat produced by the cogeneration system.

Current cogeneration systems have improved so that up to 50% of heat can be recovered, which means the heat loss is just 20%. However, to encourage the use of cogeneration systems, we must develop small cogeneration systems with high electricity generation efficiency. One possibility is a cogeneration system based on a fuel cell. Fuel cells can generate electricity at an efficiency of 50%, but the rest of the chemical energy of the fuel ends up as heat. If the fuel cell operates at a temperature of 100 °C, then the excess heat is released from the fuel cell at 100 °C. From a fuel cell operating at 100 °C and generating electricity at an efficiency of 50%, we could obtain some of the chemical energy that was not converted into electricity as hot water having a temperature of 100 °C. This hot water would contain as much as 30% of the original chemical energy. Even if we were to use some of the heat released from the fuel cell to preheat the fuel and air, there would still be an excess of heat. In fact, such a fuel cell must be equipped with a cooling system because if we did not release the heat from the fuel cell, it would overheat. In other words, even if there were no demand for heat, we would have to remove the heat from the coolant before returning it to the fuel cell. If we could develop a cogeneration system based on a fuel cell, its efficiency in generating electric power would rival that of large-scale electric power facilities, so any usable heat that is cogenerated would be an added benefit.

Reduce Energy Consumed for Heating and Cooling to One Tenth

In the last few sections, we have seen that the waste of energy resources from technologies related to space heating and cooling, refrigerators, baths, and water heating is quite large. Consequently, there should be lots of room left for reducing energy consumption. Even if we only cut the temperature difference of heat pumps by a factor of three and improve the efficiency of compression pumps from their current value of about 50% to 75%, this would still reduce the electricity consumed by air conditioners for transporting a given amount of heat to about one fifth. If we cut the heating and cooling load by half through improvement of insulation, it should be possible to reduce the electricity consumed for air conditioning to one tenth of what it was in 1995.

Refrigerators are also heat pumps. So, theoretically, it should be possible to achieve energy conservation in the same way as described for air conditioners. Furthermore, in addition to improving the insulation, we could minimize the increase in the load that occurs when the refrigerator door is opened and closed by compartmenting off the space in the refrigerator.

We could devise ways to use the waste heat from refrigerators and other appliances to heat water or provide space heating, resulting in even more energy savings.

Lighting

Lighting is a "daily life" activity with particularly low energy efficiency. Incandescent light bulbs change only 2% of electricity into light, and even fluorescent light bulbs, which we consider to be energy-saving devices, have efficiencies of only about 12%. We need to improve the efficiency of lighting devices. Semi-conductors could make an important contribution here. Special semi-conductors called light-emitting diodes are starting to appear as indicator lights for televisions and stereos, lighted road signs, and the display panels in airports and train stations. Recently, these lights have started to appear in hotels. If you see a light that you think is an LED, carefully see if it is hot. Even fluorescent light bulbs get too hot to touch. So if it isn't hot, it is probably a high-efficiency LED light. As this kind of technology develops, we should see a two to three-fold increase in efficiency in lighting, even in comparison to fluorescent light bulbs.

3 Power Plants

We have seen that technologies increasing the efficiency of electrical devices, such as air conditioners and lighting, can have a huge impact on energy use. The possibilities for conserving energy on the electricity supply side, in other words at power plants, are also great. Here, let's consider energy conservation in thermal electric power plants.

As we saw in figure 2-2, a thermal electric power plant is a mechanism to transform the chemical energy of fuel into electricity. The waste heat from a thermal power plant is the chemical energy that is not transformed into electricity. Therefore, the way to increase the efficiency of electricity generation is to minimize the part lost as heat.

The High-Temperature Limit

In the second half of the 20th century, a remarkable improvement in technologies for generating electricity in thermal power plants occurred. The electricity generation efficiency of thermal power plants, which was around 20% in the middle of the century, rose to over 40% by the end of the century. This increase in efficiency was due to technologies that made it possible to raise the temperature and pressure of the steam in the power plants. The temperature of the steam in thermal electric power plants, which was around 450 °C initially, is now over 600 °C. At the same time, the pressure of the steam, which was around 40 atmospheres initially, has increased to more than 300 atmospheres. As a result of these advances in technology, the efficiency of electricity generation could be increased to more than 42%. And currently the makers of thermal power plants are trying to push the temperature limit to 700 °C, thereby increasing the efficiency even further. However, the current temperature and pressure of the steam are close to the limits for the materials of the power plant. If we were to increase them much more, the steam would melt or corrode the iron-based materials of the turbine.

There are actually two ways to consider the efficiency of a thermal power plant. Heat is required to change water into steam, even if the temperature does not change, and when steam is changed into water, heat energy can be obtained. To calculate the efficiency of a thermal power plant, we divide the electricity obtained by the heat required to produce that electricity. However, the amount of heat available from steam depends on whether we consider the heat that is obtained when the steam providing the heat is changed into water. The amount of heat including the heat obtained when steam is changed to water is called the higher-heating value (HHV). The amount of heat obtained just when steam is lowered from the initial to final temperature is called the lower-heating value (LHV). The HHV is larger than the LHV, so the efficiency of a thermal power plant calculated in terms of the HHV will be lower than the efficiency given by the LHV. In fact, fossil fuel energy must be provided to convert water to steam in addition to raising the temperature of the steam, so the efficiency based on the HHV is probably more accurate. We will use the HHV based efficiency in this book.

Even an efficiency of 42% means that during the process of generating electricity in a thermal power plant, 58% of the chemical energy of the fuel is lost to the environment, mainly in the condenser. To make this efficiency higher, we must find a way to increase the input temperature of the turbine. The reason is that, as we saw in Chapter 2, high-temperature heat has a greater value than low-temperature heat because a larger fraction of the heat can be transformed into electricity.

Combined Cycle Electric Power Generation

The technological innovation that broke through this efficiency barrier was combining a steam turbine and a gas turbine to produce a combined cycle (figure 4-4). A gas turbine works in essentially the same way as a jet engine. In a combined cycle, first the combustion gas of the fuel is used to turn the gas turbine, and as much electricity is obtained as possible. The exhaust gas from the gas turbine still has a temperature as high as 1000 °C, so this exhaust gas is used to generate steam, and additional electricity is obtained from a normal steam turbine. The efficiency can be increased in the combined cycle because the maximum temperature at which electricity is generated is higher. Instead of being used to produce 600 °C steam, a combustion gas with a temperature as high as 1500 °C is used by the gas turbine to produce electricity directly. A commercially operated combined cycle plant with an electricity generation efficiency of 53% has been in operation since June 2007 at the Kawasaki thermal power station in Japan. Another example is GE's H system power plant in operation in Baglan Bay, Wales.

Theoretically the efficiency can be raised even further if the temperature is increased, so efforts are being made to find ways to raise the temperature of the gas turbine even higher. One problem is that the materials of gas turbines used in power plants today cannot handle temperatures much higher than 1500 °C. However, with the development of new materials and the improvement of the structural

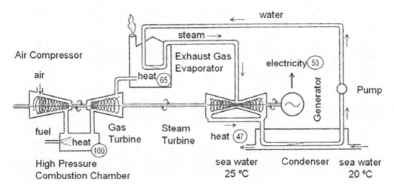

Fig. 4-4: The basic mechanism of a combined cycle gas turbine power plant
Note: The air compressor, gas turbine, and steam turbine are installed on the same axis. The system is essentially the same as the system shown in figure 2-2 with a gas turbine added. The numbers in circles are energy amounts in each of the parts when the fuel energy is 100.

design of the gas turbine, prospects look good for reaching a temperature of 1700 °C. As a result of these efforts together with advances in cooling technologies that are necessary to keep the turbine blades from deteriorating, it is thought that an electric generation efficiency of more than 55% should be possible in the near future.

Is this the limit? Not at all. The theoretical limit of electric generation efficiency is 100%. Various ways to approach this efficiency level are being studied. We just saw some ways that are being explored to make the temperature of the gas turbine even higher. Other research aimed at increasing the electric generation efficiency include devising better ways for combining the gas turbine and the steam turbine. There is even work to develop a triple stage combined cycle where before delivering the fuel to the gas turbine, electricity is first obtained from a fuel cell. Fuel cells do not consume all of the fuel that is input to the cell. The fuel that remains in the gas emitted by the fuel cell can be combusted in the gas turbine to generate more electricity. Finally, the hot exhaust gas is used to generate steam for use in the steam turbine.

In summary, the efficiencies of electricity generation using either generators or electric cells vary widely based on the methods and technologies that are used. By realizing better efficiencies, we can reduce the amount of fossil fuel we consume. This is an important part of the potential for conserving energy through technological advances.

Chapter 5
Making Things and Recycling Things

As we saw in Chapter 1, it is likely that we will face a difficult state of affairs in the 21st century, caused by the three-pronged crises of depletion of oil resources, global warming, and massive generation of wastes. In the previous two chapters, we examined the use of energy for activities in transportation, homes and offices. Clearly there is still plenty of room for improving the energy efficiency of the "daily life" activities that make up half of the total human consumption of energy. Improvements in energy efficiency help us to solve the problems of oil resource depletion and global warming. How can we address the problem of massive generation of wastes? One way is to construct a material-recycling society where waste materials are recycled into new products through the human activities of "making things." However, because one half of the energy is consumed in "making things," if recycling consumes too much energy, we will end up undoing all that we have achieved through improvements in the energy efficiency of "daily life" activities. Therefore, what we must do first is determine whether recycling with high energy efficiency is possible.

1 The Theory of Recycling

Human Artifacts Will Eventually Become Saturated in Society

We are constantly purchasing new products as old products wear out, and new buildings, roads and other infrastructure are constantly being built as cities expand. As a result human artifacts are constantly accumulating in society. This accumulation is visible in the form of our modern cities, and each new city that emerges represents a new accumulation of human artifacts. However, the earth is limited, so it is impossible for the accumulation of human artifacts to continue forever. There must be some point at which the amount of human artifacts accumulated in society levels off or "saturates." By the "saturation" of human artifacts, we mean that the amount of materials in the human artifacts disposed each year is equal to the amount of materials that are required to manufacture new artifacts. Therefore, when human artifacts become saturated in society, if we can redistribute waste materials to places where they are required through recycling, we can put an end to the exploitation of natural resources.

In fact, there are signs that the saturation of human artifacts is already happening in developed countries. For example, the car ownership in almost all developed countries is more than one car for every two people. When car ownership reaches this level, the total number of cars in society approaches saturation, and demand for new cars becomes centered on replacement buying.

In Japan, which has a population of 127 million, currently there are about 50 million passenger cars. The average time that a car is used before it is disposed is about ten years in Japan, so we can estimate that the number of new cars sold each year for replacement buying will be 5 million. Although there is some variation from year to year, following 1989, the number of new cars registered each year has in fact peaked at between 4 and 5 million. In OECD Europe and the U.S., the vehicle ownership per person increased only slightly between 2000 and 2004. Therefore, in these countries as well, the number of cars is already nearing saturation.

Construction of buildings is another example of human artifact saturation. In the large cities of Japan and Europe it is already the norm that when a new building is to be constructed, an old building must be demolished to make room for the new building. Buildings constructed on land where no building existed before are becoming the exception. In figure 5-1, we can see this state of building saturation in the annual production of cement, which is the main material for the construction of buildings. The current total global cement production is 2.5 billion tons per year. Cement production in the U.S., which used to be the world's largest producer of cement, began to saturate at around 80 million tons per year from the 1970's.

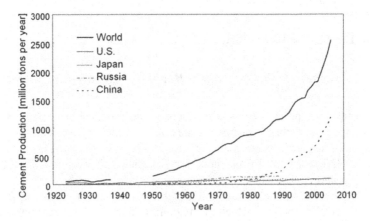

Fig. 5-1: Cement production in different countries (Data from UN Common Database, United Nations Statistics Division; and Mineral Commodity Summaries, U.S. Geological Survey)

In Japan, the amount of cement production, which grew rapidly following the war, has fluctuated between 70 million and 95 million tons per year from the second half of the 1970's, indicating a saturation of Japanese cement production. Data for Russia is limited, but it appears that Russian cement production has saturated as well.

Currently, the largest producer of cement in the world is China. China produces an astonishing 1.2 billion tons, which is almost half of the global production. Of the major cement producing countries shown in figure 5-1, China is the only country whose production has increased significantly in the past decade, and that increase has accounted for almost all of the increase in global production during that time. There is no question that if someone who visited Shanghai at the end of the 20th century were to visit the city again today, that person would be stunned by the transformation that had taken place. In the span of just a few years, what was once a sprawling rural village has become a metropolis eclipsing the modern cities of Japan, Europe and America. The population of Shanghai is 13 million, more than that of Tokyo or any city in the U.S. or Europe. Expressways and subways run through the city, and the cluster of enormous buildings bring to mind the high-rise skylines of Manhattan in New York or Shinjuku in Tokyo. One part of the 1.2 billion tons of cement that is produced in China each year continues to go into the construction of modern cities such as the new Shanghai. However, even in those cities, at some point in the future the number of buildings will approach saturation.

The Raw Material for Iron Will Inevitably Change

Let's take a look at the production of iron from this perspective of artifact saturation. It is estimated that by the end of the 20th century, humanity had produced a total of 18 billion tons of iron and that about 10 billion tons of that iron was accumulated in society as human artifacts such as cars, buildings, and bridges (some estimates are higher, but we use this conservative estimate here). In other words, most of the iron that was made through the reduction of iron ore in the past has not been thrown away as garbage or recycled, but rather has accumulated as valuable parts of the social infrastructure. So what will happen in the future to this iron?

We can estimate the rate of generation of iron scrap from the amount of iron contained in the human artifacts accumulated in society and the average lifetime of those human artifacts. The average lifetime of human artifacts made of iron is about 30 years, so one thirtieth of the iron in accumulated human artifacts appears each year as scrap. Because the current amount of iron accumulation is about 10 billion tons, more than 300 million tons of scrap is being generated each year. The amount of human artifacts accumulated in society is continuing to increase, so the amount of scrap that is generated each year will also continue to increase.

The production of iron from iron ore in 1995 was 500 million tons per year. If this production were to continue unabated, and if we also assume that all of the iron products made will be recycled as scrap and used to make other iron products, then from 1995 to 2050, more than 25 billion tons of iron will have been newly accumulated within society. At that point, the total accumulation of iron, which in 1995 was about 10 billion tons, will exceed 35 billion tons. If one thirtieth of this accumulated iron becomes scrap each year, then from 2050 1.2 billion tons of scrap will be generated each year. Therefore, in 2050, the generation of scrap alone will exceed the total iron production in 1995 of 800 million tons per year.

As a consequence, all we need to do to create a material-recycling system for iron is reduce the production of iron made from iron ore and make efficient use of the scrap instead. In Japan, the total production of iron and steel has stayed the same at about 100 million tons per year since 1980, and in 1995, 67% of production was from iron ore and 33% was from scrap. The iron and steel industry in the U.S. has a much longer history than Japan, and as a result there is a lot more accumulation of iron products in U.S. cities. A lot of scrap is generated from these products, and so the fraction of the total iron and steel that is produced from scrap in the U.S. is much higher than in Japan. In 1995, it was more than 50%.

However, as we saw in Chapter 1, the current production of iron from iron ore is about 900 millions per year, almost double the production in 1995. Does this mean that we are headed away from recycling and towards disaster? Not necessarily. The important point is that the consumption rate of iron ore is already decreasing in the developed countries, which indicates that those countries are well into a transition to a recycling society based on the use of scrap. In developing countries, the demand for new iron products is large, so the production of iron from the reduction of iron ore will most likely continue for awhile. Most of the recent increase in production of iron from iron ore has occurred in China and India. However, eventually even those countries will move towards the same form of scrap-based recycling as the developed countries.

The concepts of recycling presented above are not limited to iron – the same thing can be said for other types of material production. Figure 5-2 gives a sketch of the transition of production that is necessary for achieving a sustainable society. Where we are on the horizontal axis depends on the material considered and the level of development of the country. For most materials considered on a global level, as a result of continued demand for new human artifacts in the near future, the accumulation of human artifacts will increase, and the generation of waste will also increase proportionally. However, by increasing the annual production of materials from recycling waste artifacts, we will begin to reduce the consumption of non-renewable natural resources. In this way, we should be able to circumvent the problem of the exhaustibility of non-renewable natural resources. Therefore, the real problem that we must address is the future of energy resources.

Let's think a bit more about the conclusion in the previous paragraph. It is often said that we must break away from our mass production / mass consumption civilization. However, we should consider carefully what this means. In order to meet the basic material demands of the more than six billion people living on the earth, we cannot avoid the need for producing a huge amount of materials. On the other hand, we have seen that the major threats to the sustainability of human society are

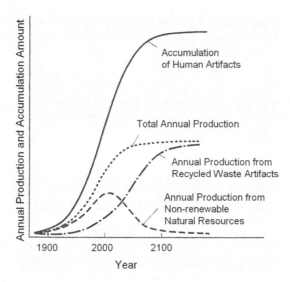

Fig. 5-2: A graph showing how the accumulation and production of human artifacts will progress in the 21st century

the depletion of resources, particularly oil, and the massive generation of wastes. It is not mass production itself that is the problem. Therefore, to achieve sustainability what we must aim for is to break away from a civilization based on massive consumption of the earth's natural resources and massive dumping of waste materials into the earth's environment. The warnings of scientists and other experts that the amount of resources and the capacity of the environment are limited and that human activities are already exceeding those limits are important. However, there is no need to despair. There is a solution.

2 Recycling That Is Also Energy Conservation

Many people have expressed negative opinions regarding recycling. Among them is the objection that if we recycle, we will use too much energy. Let's examine this criticism using the method of breaking down processes into elementary steps that we developed in Chapter 3.

The collection of scrap and other waste material is basically "transportation," and so theoretically the energy required for collection is zero. Of course, in reality we cannot avoid having to consume some energy to collect the waste material. However, for production from natural resources, we need to extract and transport raw materials from mines that often are in remote locations. At least in energy terms, in most cases the collection required for recycling is not much greater than the transport required for production from natural resources.

Some energy is required for the "separation" process of obtaining basic materials from mixtures of waste material. However, as we saw in Chapter 3, the relative size of that energy is 1, and so if we do the separation efficiently, it will not require a lot of energy. Therefore, the real problem is how much energy is consumed at the plant in the process of producing new basic materials from waste materials that have been collected and separated. In the next several sections, we will compare production from natural resources and from waste human artifacts for iron, aluminum, cement, and glass.

Reduction of Iron Ore: The Blast Furnace Method

We saw in Chapter 1 that of the 1.3 billion tons of iron produced each year, most is made in blast furnaces using iron ore as the raw material, but a significant fraction is made in electric arc furnaces using scrap as the raw material.

First, let's break down the production from iron ore into elementary steps. The production process occurs via the following three steps. First, the reduction reaction uses carbon to change iron ore into pig iron and CO_2. Next, the carbon contained in the pig iron is separated out and the concentrations of trace elements in the iron are adjusted as required by the product specifications. Finally, the iron material is shaped into iron products such as thin sheets and rounded bars. Therefore, the process consists of the steps reduction, separation, and shaping. Theoretically, the energy needed for shaping is zero, and in fact through advances in integrated iron and steel making such as the continuous casting process we will see next, the energy used to make iron into sheets and bars has been reduced dramatically. So most of the energy required for producing iron from iron ore is used in the reduction and separation steps. The relative size of the energy requirement is 1000 for reduction and 1 for separation. Therefore, if performed efficiently, the energy requirement for separation is negligible.

By calculating the minimum energy needed for the reduction of iron and comparing it to the maximum energy that can be obtained from the combustion of carbon, we can find the minimum amount of carbon that is required for the reduction step. The theoretical minimum energy needed for making iron converts to 202 kg of carbon for the manufacture of one ton of iron. Currently, the value for large-scale integrated iron and steel works is about 600 kg. Therefore, we see that one third of the carbon is necessary for making iron even in the ideal case, and only two thirds of the carbon can be saved by even the most sophisticated technologies.

In the past, after the pig iron came out in molten form from the blast furnace, it was cooled into blocks for storage. The blocks of iron were later heated and shaped into thick plates, which were then left to cool once more. This process of reheating and cooling the iron was repeated until little by little the desired shape, such as a thin sheet that could be used for the body of a car, was obtained. The energy used to heat the iron each time was not recovered, so a large amount of fuel was consumed. To reduce this waste of fuel, the continuous casting process was introduced. In continuous casting, the steps from the production of pig iron at the blast furnace to the forming of iron sheets and bars are carried out in a continuous process so as to avoid repeated heating and cooling of the iron. Furthermore, other technologies were developed to recover some of the energy that was input as coke into the iron making process. One is to use the gas emitted from the blast furnace, which contains fuel such as carbon monoxide and hydrogen, to generate electricity. Another is to generate electricity using devices like top-pressure recovery turbines where a turbine for generating electricity is turned by the pressure of the exhaust gas. Through the application of these various technologies, the high present day overall efficiency of iron and steel making – 600 kg of coal per ton of iron – has been achieved.

However, currently almost none of the heat that is used for heating the iron ore in the blast furnace is recovered. Also, there are many steps in the process of shaping and forming, such as rolling and cutting, where energy is still wasted. We have seen that the energy needed for shaping and forming theoretically is zero, so any energy used for that purpose is waste. For example, in one process a slab of iron as much as one meter thick is rolled into thin sheets having a thickness of 0.7 mm for use in making automobile parts. Currently, this process is carried out by passing the iron slab through a large number of rollers that are powered by electricity. Even though each step of rolling may not consume so much energy, when the multiple steps are added up, the total energy consumption is rather large. Also, energy is used for processes such as surface treatment.

The difference between the ideal carbon consumption rate of 202 kg per ton of iron and the actual value of 600 kg is the result of the combination of these various small energy consuming steps. How much further conservation of energy will be achieved in new iron making facilities will depend on how much is invested in equipment for that purpose; however, it is unlikely that we will be able to reduce the carbon consumption to less than 400 kg per ton of iron in the foreseeable future.

Recycling of Iron: The Electric Arc Furnace Method

Iron is recycled in the following way. Iron products that have reached the end of their lifetimes are dismantled, and the iron is sorted from the other materials and collected as scrap. The scrap is melted, impurities are separated out, and the iron is shaped again and shipped out as iron products such as rods and sheets. The furnace where the scrap is melted, called an electric arc furnace, uses electricity to generate the heat for melting the scrap.

Dividing this process of making iron from recycled scrap into elementary steps, we have melting, separation, and shaping. Of these steps, the ones that require energy are melting and separation, and the sizes are 10 and 1, respectively. Therefore, the largest part is melting. Converting the heat that is required to melt iron into units of carbon, we find that 7.5 kg of carbon is sufficient to melt one ton of iron. This is about 27 times less than the ideal minimum value of 202 kg for the reduction of iron ore, so we see that recycling iron has the potential to be much more energy efficient than producing iron from natural ore.

In reality, recycling of iron today is not so efficient. The electric arc furnace melts the iron scrap by converting electricity directly into heat, which we have seen is an inefficient use of electricity. Furthermore, as with the processes that we looked at in Chapter 3, in order to melt the iron scrap quickly, the temperature of the electric arc furnace is made much higher than would be required in the ideal case. When we calculate the fuel consumed at a thermal power plant to generate the electricity that is currently used in iron scrap recycling, we find that 300 kg of fossil fuels are actually consumed for each ton of iron scrap. Still, this is just half of the 600 kg used in the blast furnace method, so for the manufacture of iron, even recycling using this rather inefficient method consumes much less energy than production from natural resources.

Let's summarize the points above. The blast furnace method of making iron from natural resources requires energy for the reduction of iron ore. The electric arc furnace method for recycling iron from iron scrap requires energy for melting. The sizes of the corresponding elementary steps of reduction and melting are 1000 and 10, respectively, so we can estimate roughly that the energy consumption for the electric arc method should be on the order of 100 times smaller. In fact, we saw that the melting heat of iron is about one twenty-seventh the heat of reduction. This is the basis for the energy savings of the electric arc furnace method. However, given that the melting heat is just one twenty-seventh the heat of reduction, why is it that conventional iron scrap recycling can only reduce energy consumption by half that used by the blast furnace?

One reason conventional iron scrap recycling does not achieve a higher reduction in energy consumption is that almost none of the heat energy generated in an electric arc furnace to melt the iron scrap is collected. As one example, this energy could be used to replace the electricity that is now consumed for the shaping and forming of the iron. However, the most important problem is that the heat for melting the iron scrap is currently supplied using electricity. The reason is that using electricity it is easy to obtain the high temperature of 1540 °C that is required for melting iron. However, combusting fossil fuels, converting about 40% of that heat into electricity using a thermal power plant, and then changing the electricity back into heat to melt the iron is terribly inefficient, as we saw in the comparison of using an electric heater versus a gas stove to heat a room. It is possible to develop technologies to melt iron scrap using fossil fuels directly, and researchers are currently working on practical applications. By using fossil fuels instead of electricity to melt the iron scrap, it should be possible to reduce the energy consumption of the electric arc furnace method by 50%, or 150 kg of carbon per ton of iron.

Electrolysis Versus Electric Melting of Aluminum

Next, let's take a look at aluminum. If we look at the production process from bauxite in terms of elementary steps, we have mining that is a form of transportation, melting of bauxite, electrolysis of the bauxite that is a form of reduction, and shaping. The steps that require energy are melting and electrolysis; however, because the sizes are 10 and 1000 respectively, we can see that most of the energy is consumed as electricity in the electrolysis of the bauxite. Currently, the electrolysis process is carried out at a voltage that is about twice the theoretical value, so about twice the ideal amount of electricity is consumed. Although the electricity for electrolysis of aluminum is usually provided by hydropower, even hydropower loses 15% of the potential energy of the hydropower resources in generating electricity. Therefore, the energy conservation potential is almost 60%.

The recycling of aluminum is quite widespread. The reason is that, like iron, the consumption of energy for recycling aluminum is small, and therefore it is sufficiently cost-effective to recycle aluminum even in pure economic terms. The heat of melting for aluminum is about 83 times less than the heat of reduction required for electrolysis of bauxite, and even in actual industrial applications, the electricity used in plants for aluminum remelting and rolling is no more than 3% that used for production from bauxite. Therefore, the energy-related benefit of recycling is even larger for aluminum than it is for iron in both theoretical and practical terms.

Recycling of Non-metal Mineral Materials

Looking in the same way at the process of cement manufacture, we see that it is made up of the following elementary steps: mining of limestone that is a form of transportation, pulverization that is a form of shaping, and the reaction of thermal decomposition that removes CO_2 from limestone to produce calcium oxide. Theoretically, other than the reaction, none of the steps need to consume energy. Furthermore, compared to the reduction step with an energy measure of 1000 that is required in the manufacture of metals such as iron and aluminum, the energy measure for reactions is just 100, so we can estimate that the theoretical energy consumption for making cement is about one tenth that required for metal. In reality, production of one ton of cement only requires 100 kg of fossil fuel, which is six times less than the amount used for iron making. Furthermore, this value is just 40% larger than the theoretical value of energy required to make cement, which is about 70 kg.

There are many types of cement. Normal cement, called "Ordinary Portland Cement," can be made up of as much as 5% materials from other processes, such as the byproduct of blast furnaces called "blast furnace slag," the residuals from combustion of coal called "fly ash," and even ordinary limestone. Another type of cement, called "Portland Cement Blends," is characterized by a larger amount of additives. The different types are used for different purposes. In this way, waste materials from other processes are recycled as much as possible in the production of cement.

The other main non-metal mineral-based material, glass, is produced through the following steps: 1) mining of the raw materials silicon dioxide from quartz, calcium carbonate from limestone, and sodium carbonate from soda ash, 2) pulverization, 3) mixing, 4) melting, 5) thermal decomposition, the same reaction used in making cement, 6) melting, and 7) shaping. Mixing is the opposite of separation, and so because separation requires energy, we know that mixing is an energy producing process. Therefore, the only steps that require energy are melting and reaction, with sizes of 10 and 100, respectively. However, while the reaction only involves calcium carbonate, the melting process must be done for all of the materials, so the energy consumption for melting cannot be ignored. Currently, 200 kg of fossil fuels is used to produce one ton of glass. This is more than three times larger than the theoretical energy required for both the melting and the reaction, which corresponds to 60 kg of fossil fuels per ton of glass.

Why is the ratio between the current energy consumption rate and the theoretical value so different for glass and cement, if their manufacturing processes are almost identical? The main reason is the difference in the quality requirements of the products. Glass products have strict requirements for quality. For example, contamination by even a small amount of bubbles or other impurities cannot be allowed. Therefore, the manufacturing process must be carried out slowly and carefully. For that reason, the glass material needs to be kept hot for a longer time than the cement material, and this means a larger heat loss in actual production processes.

Currently, about 50% of glass is recycled. Although not to the extent of the recycling of aluminum and iron, the energy consumption of production from pulverized recycled glass, called "cullet," is smaller than from natural materials. Therefore, like we have seen in the manufacture of other materials, recycling of glass is advantageous from an energy perspective.

Almost all of the cement that is produced in the world today is mixed with sand, gravel and water and used as concrete. As we saw in Chapter 1, after the concrete products reach the end of their product lives, the concrete is recycled by pulverizing it and using it in low-grade applications such as roadbeds. However, we also saw that in the future, this kind of demand will begin to decrease. Therefore, there will be a need for a full-fledged cement recycling process where cement is remade from the waste concrete produced, for example, during the demolition of a building. If we pulverize the concrete, separate out the sand and gravel, and heat the remaining material, which is calcium oxide hydrate, it is possible to recover the cement. The theoretical energy size for pulverization, separation and reaction is 0, 1, and 100 respectively. This is the same as the breakdown that we saw for the current cement production process. Therefore, technologically it should be possible to develop a recycling process that can be operated with the same level of energy consumption as the current cement production process. In the future, we may see pulverizing mixer trucks that can recover cement from concrete on site in place of concrete mixer trucks.

Recycling Is a Means for Energy Conservation

We have seen in the previous sections that there are still numerous possibilities for conserving energy in manufacture of metals, cement, and glass from natural materials. However, more importantly, we have also seen that the processes of separation and melting for recycling these materials from waste products actually consume less energy than the processes required for manufacture from natural materials. The difference is particularly large in the case of metals. Furthermore, we have seen that this is not only true in terms of the theoretical energy consumption required of all of the steps from collection to reuse, it is also true in actual recycling applications for metals and glass. Therefore, in most cases the criticism that recycling results in the waste of energy is just not true.

If we do come across a recycling process that results in a large consumption of energy, we should consider this to be an indication of large inefficiencies in the process. Just as we saw in the example of heating a room, the energy efficiency for recycling is strongly affected by the method that is used. For instance, if the waste material that is collected is a mixture of all kinds of substances jumbled together, consumption of a large amount of energy is probably unavoidable in order to recycle that material.

We saw in Chapter 1 that the amount of elements in the biosphere is constant. So what does it mean for a resource such as iron to become depleted? As we saw with energy in Chapter 2, the key is in what it means to be a valuable resource. The conditions for a potential resource, natural or manmade, to be valuable in terms of energy are as follows. First, the resource must have a high concentration of some basic material. Second, the resource must not contain too many elements that are difficult to separate. Third, the resource must exist in large amounts that are gathered together. Without these conditions, because elements exist throughout the biosphere, there would be no limit to the amount of available resources. For example, large amounts of almost all kinds of elements are contained in sea water, including metals and even uranium. However, because the concentration is extremely low, a huge amount of energy is needed to extract these elements from sea water. Therefore, as a resource, sea water cannot compete with mineral deposits under ground.

We can consider the recovery of materials from human artifacts that have been thrown away in the same manner. The first condition states that in order for waste products to be valuable resources, the concentration must not be significantly less than that of natural resources. Clearly the concentration of materials such as iron, glass and paper in waste products is not lower than in natural resources. The real problems are related to the second and third conditions: waste products contain elements that are difficult to separate, and waste products are generated in small amounts all throughout society. Therefore, there are two key points to raising the efficiency of recycling and the value of waste products as a resource. First, we must make sure that when products are thrown away, materials that are difficult to separate are not mixed in. Second, we must construct a system for efficiently collecting and transporting the waste products that are spread out in society's infrastructure.

Moreover, if we can succeed in constructing a comprehensive system based on the concepts we have seen here that encompasses the entire lifecycle of human artifacts from design to consumption and disposal, this will help us to reach our overall goal of realizing a material-recycling society with superb energy efficiency. We will come back to the issues related to realizing this system in Chapter 8.

3 Is It Bad to Burn Waste Paper and Plastic?

Even if We Burn Paper, It Can Still Be Recycled

We have seen here that recycling gives us the twin benefits of 1) reducing the amount of waste materials that get dumped in the biosphere and 2) conserving energy. However, we should keep in mind that when we recycle things, it is not always necessary that they be circulated as materials. This is important for two reasons. The first is that unavoidably some waste material will be generated whose quality is too degraded to be easily reused as a raw material for producing new material. The second is that we have a large need for energy. These considerations are particularly important for paper and plastic. First, let us consider the case of paper.

From the direct mail advertisements that bombard our mailboxes to the recent deluge of computer printouts, a huge amount of paper that seems almost criminal to throw away is being consumed each day. We saw in Chapter 1 that the production of paper from trees consumes a lot of energy. Like metals and glass, recycling paper if done efficiently can save energy. Currently, in Japan, the EU and the U.S., about 50% of waste paper is used together with new pulp in the production of paper. If we can increase this recycle ratio without stretching technological and economic limits, then it is desirable to do so. However, if we repeat the recycling of paper over and over, gradually the length of the fibers will become shorter, and the strength of the paper will decline. For this reason, the limit for the ratio of waste paper in the raw material for paper manufacture is said to be 70%. If we cannot recycle all of the waste paper directly due to this reason, what is the best alternative?

Currently, waste paper that is not recycled is incinerated together with municipal waste or simply buried in landfills. The paper buried in landfills decays or is consumed by microbes. Therefore, whether the waste paper is incinerated or buried in land fills, it eventually ends up as CO_2 in the atmosphere.

If we are going to burn the waste paper anyway, we should try to find a useful way to burn it. Just disposing the waste paper in incinerators or land fills is the same as "burning oil fields." However, if we burn the paper in a coal-fired power plant, we can reduce the amount of coal consumption by the amount of heat that is generated by the paper. Using waste paper in cement making plants or blast furnaces is also possible. Wherever fossil fuels are burned, if we can substitute waste paper for some of the fossil fuels, we can reduce the use of fossil fuel resources. The question we should ask is not whether or not burning is wasteful, but rather what is best in comparison to the current situation of waste paper disposal in incinerators or land fills.

In Chapter 2, we saw how the efficiency of heating depends greatly on the method that is used. When we use waste paper as a fuel, we must also consider what method will give the best efficiency.

For example, refuse power generation is one technology that is used for recycling garbage. The idea is to burn garbage and to use the heat for thermal power generation. Unfortunately, the power generation efficiency that can be achieved is little more than 10%, just a fifth of the efficiency of the most advanced power plants. In other words, fuel in normal power plants can be used five times more efficiently than in refuse power generation. Refuse power generation is also used to produce hot water. However, as we saw with cogeneration in the last chapter, in most cases the demand for hot water is much less than for electricity. So even if we can collect almost all of the heat from the combustion of garbage in the form of hot water, the value of that energy will be low.

If we can find a way to burn waste paper that reduces consumption of an amount of fossil fuels equivalent to the chemical energy of the paper, then it is probably alright to burn the paper. This may require us to find a way to efficiently remove water and other contaminants from the waste paper. However, if we can do this without using too much energy, all of the energy that we can obtain from burning the paper will be a positive effect in terms of depletion of fossil fuel resources.

So why is it that we feel burning paper is wasteful? One reason is probably our concern that consuming paper results in the destruction of forests. However, if we are careful in managing the forests and replanting the trees in a sustainable way, then burning paper to produce electricity, for example, can actually be considered as a form of natural solar-powered energy system. The other important reason for our resistance to the idea of burning paper is our lack of recognition that in reality we are already burning an amount of oil that is more than ten times the amount of paper we use. For example, in comparison to the 2.7 tons of fossil fuels in carbon units that Japanese people use per person each year, the amount of paper use is just a little more than 0.2 tons. There is no reason that we must not burn waste paper at the end of its lifecycle. What we must do in order to make the production of paper sustainable is replant the trees after they are harvested for making pulp and reduce the current amount of 300 kg of fossil fuels that are burned in the manufacture of one ton of paper. Furthermore, although not treated in this book, we must also address the problem of consumption – do we really need to use this much paper?

Using Plastic as Fuel

We can use the same kind of thinking when we consider the optimal way to recycle plastic. Currently, the largest natural resource consumed by humans is the 7.5

billion tons of fossil fuels in carbon units each year that we saw in Chapter 2. Of this, the amount that is made into materials is just the 200 million tons of plastics, synthetic fibers and other petrochemical materials manufactured each year. Therefore, 7.3 billion tons of fossil fuels or almost 98% of the total consumption are burned to provide energy for "making things" and "daily life" activities.

We can divide the methods for using waste plastic into four basic types: 1) reuse of the waste plastic as is, 2) reuse after reshaping, 3) use of thermal decomposition to transform the waste plastic back into its raw material form such as ethylene, and 4) use of the waste plastic as fuel. If it is possible to reuse a plastic product as is or reshape it into a recycled product having about the same value as the original product, then that is probably the best thing to do. The energy for shaping is small, so even if we have to reshape the plastic into new products, this would still let us save nearly all of the one ton of oil consumed per ton of plastic when made from naphtha.

For the third type of recycling, where thermal decomposition is used to return the plastic to raw material form, we must be careful to consider the size of the energy consumption that would be required. As we saw at the end of Chapter 3, thermal decomposition is also the most energy intensive step in the production of plastic from naphtha. Therefore, it is not impossible that in the worst case more than one ton of oil will be consumed to recycle a ton of plastic. Also, we must take care in situations where high quality products are reused to make low quality items such as park benches and planters. If the waste plastic is reused in products that are actually needed, then it may be alright to do this. However, in some situations today recycling is done for its own sake with little consideration of how much demand there will be for the recycled products, and in other situations the recycled products are products that could have been made through the consumption of less resources if a different material was used. In these situations, it may be better to use the waste plastic as a fuel to substitute for fossil fuel resources.

For example, currently one of the most promising methods to recycle waste plastic is to use it as a substitute for coke in the reduction of iron ore. If plastic is preprocessed to remove chlorine and other impurities and then heat-treated, we can obtain grains of carbon that have almost the same characteristics as coke made from coal. Even with the technology available today, it is said that 70% of the chemical energy of waste plastic can be reused as a substitute for coke, which is excellent performance for a recycling process.

From the previous discussion, it is clear that particularly for paper and plastic, we need to consider the pros and cons of different options for recycling and reuse of waste materials from a global perspective rather than just from a single aspect such as whether or not waste products are recycled into other material products.

In a society where human artifacts have reached saturation, there are two paths for the human artifacts that have reached the end of their product lives: they can be thrown away or they can be recycled. We can imagine what would happen if we choose to throw human artifacts away by thinking about the fate of modern cities. We have seen that cities represent the accumulation of human artifacts. If we take the average lifespan of human artifacts to be 50 years, then after 50 years, an amount of waste material equivalent to all of the cities that exist today will have to be disposed of somewhere in the biosphere. If the number of cities continues to grow, and those cities are also disposed of every 50 years, then the earth will end up being turned into a garbage dump. Therefore, if we want to achieve a sustainable earth, there is no alternative but for us to work to create a material-recycling society. The point of this chapter is that not only is "making things" by recycling possible, it can also contribute to the conservation of energy resources.

Chapter 6
Introduction of Renewable Energy

The previous chapters have shown that there is considerable potential for energy conservation in the activities of "daily life." Furthermore, even for the activities of "making things," we can save energy resources through recycling in comparison to the present practice of production from natural resources. However, even if we can reduce the amount of energy that we consume in this way, we will still need a large amount of energy resources. We cannot continue to depend on fossil fuels. If we just consider the single issue of global warming caused by CO_2 emissions, it is clear that we do not have much time left to develop energy resources that can replace fossil fuels.

We have seen that it will probably not be possible to achieve the complete replacement of fossil fuels within the 21st century. However, this does not mean that we can just sit back and do nothing as we watch fossil fuel resources disappear. Rather, we must see this as a warning that only if we apply our best efforts towards the development of alternative energy resources now will it be possible for us to launch ourselves away from oil and other fossil fuels and make a soft landing to an alternative and sustainable energy system.

1 Could Intensification of Nuclear Power Be the Answer?

As we saw in Chapter 2, the options for alternative energy are limited to nuclear energy and renewable energy. To which of these should we entrust our future?

Types of Nuclear Power

Many experts claim that nuclear power is the answer. One benefit is that, because the nuclear reaction of uranium is used instead of the combustion of carbon, nuclear power causes essentially no greenhouse gas emissions. On the other hand, like fossil fuels, uranium is a non-renewable resource. While uranium does exist in rather large quantities under the earth's surface, most of it is Uranium 238, which cannot be used directly in nuclear fission. Only 0.7% of the uranium on the earth is Uranium 235, the fissionable form of uranium that can be used in conventional nuclear reactors. The amount of confirmed Uranium 235 reserves divided by the current production rate is currently just 45 years, which gives us some concern that the natural uranium resources may be exhausted. However, it is also said that if we look we can find all that we need. From the example of oil in the past, at least we can say that it is unlikely that the resources will actually be depleted in 45 years. However, this does not change the fact that current nuclear power generation is a technology that relies on a non-renewable resource.

One possible solution to this problem that has generated much interest is the use of breeder reactors. Currently, the concentration of Uranium 235 in the uranium needs to be enriched to about 2% for use as the fuel in nuclear power generation. The Uranium 238 is unused and must be disposed of in expensive containment facilities. However, if breeder reactors can be realized, it will be possible to transform the unreacted Uranium 238 that remains in the reactor into Plutonium 239, which is another fissionable material, by bombarding it with neutrons. All at once, the amount of nuclear power resources could be increased ten fold. This may seem like a perfect technology; however, unfortunately it is not without problems. Plutonium is even more dangerous than uranium, so the safety and non-proliferation issues are even more severe in the case of breeder reactors.

For a long time, people have hoped to develop a technology for producing electricity through the process of nuclear fusion. Production of electricity through nuclear fusion would work by the same principle as that which gives the sun its energy, so scientifically it should certainly be possible. If power generation through nuclear fusion could be realized, the amount of electricity that could be produced would be essentially limitless. However, considering that as of yet no one has been able to reach the critical state where the energy that is produced is greater than the energy that is supplied, and that people who were saying thirty years ago that "in thirty years we will construct a demonstration reactor" are still saying the same thing today, nuclear fusion will probably not be a viable energy source for the 21st century. If we are going to use nuclear energy, it will most likely have to be nuclear fission, with all of its resource, safety and nuclear proliferation related problems.

Concerns About Safety

Concerns regarding the safety of nuclear technology are numerous. While some of the fears may actually be unfounded, many of them are quite serious, such as the issue of nuclear weapon proliferation and the disposal of radioactive waste having a half-life of several thousand years. The contribution of nuclear power to the total global energy production is currently 5%, and it is not likely to increase much. If, for instance, we wanted to meet the total energy used today with nuclear power, that will mean constructing ten thousand plants the size of the Three Mile Island nuclear power plant around the world. The task of figuring out how to solve the issues related to accidents, terrorism and handling of radioactive waste would almost certainly exceed our current abilities.

If we cannot place our expectations on intensification of nuclear energy, we will need to focus our efforts into the development of renewable energy. Renewable energy exists in great abundance throughout the biosphere; the problem that renewable energy technologies attempt to address is how to transform that energy into forms that are easy to use, such as electricity and vehicle fuel. Numerous types of renewable energy technologies that have been proposed, ranging from solar heating and wind turbines to methods for generating electricity using the temperature difference created by the sun between the surface and deep waters of the ocean or using the osmotic pressure between salt water and fresh water that we saw in Chapter 2. However, here we will restrict our attention to those technologies that could be introduced at a significant scale in the near future.

2 Sunlight

Sufficient Amount and Excellent Quality

We can calculate the total amount of sunlight that shines down on the earth by multiplying the intensity of the solar irradiation outside the atmosphere that is directed perpendicular to the surface of the earth (which is $1.37\,\text{kW/m}^2$) by the cross-sectional area of the earth. This value is on the order of 10,000 times the total amount of energy that is used by humanity today, so the amount of sunlight energy is more than sufficient. The next problems that we must consider when using sunlight as an energy resource are its quality and density.

What is the quality of sunlight? We saw in Chapter 2 that all kinds of energy except for heat can at least theoretically be transformed with 100% efficiency and thus have the same value or quality. More accurately, all kinds of energy have equal quality except for the kinetic energy of randomly vibrating molecules that is the heat embodied in an object and the radiant energy that is produced by an object at high temperature such as the filament of a light bulb. Sunlight is radiant energy that is produced by the sun, so its quality is not as high as the other kinds of energy that we looked at in Chapter 2, such as electricity and work. Let's consider the quality of sunlight from two viewpoints.

The first is the temperature of the energy of sunlight if it is converted into heat. We saw that the value of heat is given by the temperature difference with the environment divided by the temperature of the heat, so the higher the temperature of the heat the higher its value is. The surface temperature of the sun is about 6000 °C, so sunlight has an energetic value equivalent to heat with a temperature of 6000 °C. Using the environment temperature of the earth, which is about 15 °C, we find that the temperature difference divided by the temperature is about 0.95. This means sunlight energy can be changed into electricity or work with 95% efficiency, so sunlight is energy having nearly the same quality as electricity.

The other way to think about the quality of sunlight is in terms of its wavelength. When sunlight passes through a prism or a drop of water, we see all of the colors of the rainbow. Sunlight is made up of a lot of electromagnetic waves having different wave lengths, each of which produces a different color of the rainbow. Because things like prisms and water droplets bend light to different degrees depending on the wavelength, sunlight can be divided up into different colors, as shown in figure 6-1. The wave lengths of visible light, the colors of the rainbow that we can see with the naked eye, range from 0.7 microns for red light, which is the longest, to 0.4 microns for violet light, which is the shortest. Therefore, visible light is made up of electromagnetic waves having wavelengths between 0.4 and 0.7 microns.

However, there are electromagnetic waves outside of the colors of the rainbow that exist in sunlight even though they cannot be seen by the human eye. The part with a wavelength longer than red light, more than 0.7 microns, is called "infrared radiation," and those electromagnetic waves exist outside the red edge of the rainbow. The part with a wavelength shorter than violet light, less than 0.4 microns, is called "ultraviolet radiation," and those electromagnetic waves exist outside the violet edge of the rainbow. The fraction of energy contained in each of the parts of sunlight shining on the earth from outer space is 9% for ultraviolet radiation, 47% for visible light, and 44% for infrared radiation. Ultraviolet radiation is absorbed by the ozone layer in the stratosphere, so just a tiny amount of that part reaches the earth's surface.

The energetic quality of light, which can be thought of as a flow of energetic particles called "photons," is determined by the wavelength. We can think of light with a short wavelength as the flow of particles of light having large amounts of energy, and light with a long wavelength as the flow of particles of light having small amounts of energy. For example, no matter how long you expose yourself to

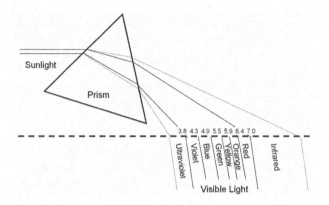

Fig. 6-1: The wavelengths of sunlight

infrared radiation, you will not get a sun tan. The reason is that the energy of one photon of infrared radiation is not enough to drive the chemical reaction of melanin that causes your skin to tan. If you stand in front of a hot stove or electric heater for a long time, you may get burned, but you will not get tanned. In order to cause the tanning reaction, the energy of ultraviolet photons is necessary. Likewise, photons having at least the energy of visible light are necessary to cause the reaction to split water; it is impossible to do with infrared radiation. Furthermore, as we might expect, visible light photons are necessary to drive the reactions of photosynthesis, and infrared radiation is not enough. That is why plants cannot grow in a room with no visible light, even if there is a strong source of infrared radiation such as a heat lamp. Finally, the wavelength of light also determines the maximum voltage at which electricity can be generated by a solar cell. With visible light, it is possible to generate electricity with more than 1.5 volts.

In summary, visible sunlight can cause the splitting of water or the reactions of photosynthesis, and with it we can make solar cells that have voltage sufficient for meeting electric power needs. Because almost half of sunlight energy is in the form of visible light, sunlight is clearly a high quality energy resource that can be used for a wide range of energy needs.

The Maximum Power of a Solar Car Is Two Horsepowers

The main problem with sunlight is its low density. As we saw in the previous section, the sunlight intensity outside the atmosphere is 1.37 kW per square meter; however, about 30% of that energy is reflected by clouds and dust and does not reach the earth's surface. When we add in the effects of the seasons, day and night, weather, and so on, the energy density of sunlight in Japan for example is no more than 200 W per square meter.

Can we make a car that runs on just solar cells? If we could, we would go a long way towards alleviating the energy resource problem. In fact, there is a solar car race that has been held since 1987, which gives us reason to hope. However, even the winners of the race cannot produce the horse-power required for regular driving conditions. If we cover a large car from roof to hood with solar cells having an electricity conversion efficiency of 15% such as those that are currently on the market, under the most intense solar irradiation at noon on a mid-summer's day, we can get about two horse-powers of propulsion force, and for average solar irradiation, we can only produce 0.4 horse-powers. Furthermore, under some weather conditions, such as cloudy or rainy days, the power level is even lower, and of course at night almost no power can be produced by the solar cells at all. Compared to the 100 horse-power engines of conventional automobiles, even under the best conditions, solar cars cannot provide enough power. In the solar car race, thin vehicles are made from light-weight materials, and solar cells are mounted on large wing-like structures on the vehicle. Even so the vehicles seem to move at a leisurely pace across the race track.

Prospects for the commercialization of solar cars are slim, solar powered commercial airplanes are nearly inconceivable, and even stationary solar cell power plants are difficult to construct because they require such a large area. All of these problems result from the low density of sunlight together with rapid fluctuations in time due to clouds and other factors. These are the main drawbacks of sunlight when looked at as an energy resource. In order to use sunlight as a source of energy, we need a large area to gather the energy and a way to store it for when the solar irradiation is weak. Two technologies that show particular promise for overcoming these kinds of problems are biomass and solar powered electricity generation.

We have seen that the theoretical maximum efficiency for converting sunlight into electricity or work is 95%. Because low density is the main problem with sunlight, we should try to get as close as possible to this theoretical efficiency in order to reduce the area required for collection. In the next sections, let's see what kind of efficiency can actually be obtained using biomass and solar powered electricity generation technologies.

Biomass Is 5%

Sunlight is absorbed by special bodies in plant cells called chloroplasts, and the absorbed sunlight gives its energy to the electrons in the chloroplasts. Photosynthesis is the process of using those electrons to synthesize fructose from CO_2 and water, and it occurs through many steps including dozens of enzymatic and ionic reactions. Fructose is a kind of carbohydrate, a chemical compound of carbon and water. Energetically, it is close to carbon, which means that its chemical energy content is comparable to coal. The efficiency of photosynthesis is high in the sense that all of the electrons that have absorbed sunlight are used. However, this does not mean that all of the energy of the solar irradiation can be used. There are two main reasons, and the essence of both is that, as we saw before, sunlight is composed of light with different wavelengths.

The first reason is that, as was noted earlier in this chapter, the energy of infrared photons is too small to be absorbed by the chloroplasts in plants, so about 44% of the energy of sunlight cannot be used for photosynthesis. The second reason is that chloroplasts can only make use of the energy in a photon that is equivalent to that of a photon of red light. The excess energy of photons of light that is more energetic than red light, such as blue and violet, ends up becoming heat. In short, chloroplasts can only absorb photons of sunlight with wavelengths within the range of visible light, and furthermore, red is the only wavelength of light for which the process of photosynthesis is optimal.

Simply as a result of the suboptimal efficiency for using the energy of wavelengths of sunlight other than red light and the inability to use infrared light at all, the maximum efficiency of photosynthesis drops to less than 40%. Moreover, through energy losses during the many reaction steps leading to the production of fructose, efficiency drops further to about 10%. Furthermore, not all of the visible light in sunlight even reaches the chloroplasts in plants. If leaves absorbed all visible light then they would appear black, but in fact they appear green, which means that green light is reflected. Combined with several other factors that lower the efficiency of photosynthesis, we find that the theoretical efficiency limit for photosynthesis is only about 8%.

The carbohydrates produced from sunlight by photosynthesis are accumulated in the body of the plant as "biomass." Plants consume about half of the accumulated biomass themselves through respiration. Furthermore, plants only grow from spring to summer, lying dormant in autumn and winter. After all of these factors are taken into account, the maximum efficiency of biomass in the sense of the fraction of the year long solar irradiation energy that is available as harvestable biomass for human use ends up being less than 1%.

For example, rice is a crop that uses sunlight with relatively high efficiency. Thin leaves and stalks grow together densely, so that nearly all of the sunlight shining down on the rice field is collected. Rice has a high crop yield of about 10 tons per hectare, and if we include leaves and stalks, about 20 tons of biomass can be harvested. The overall efficiency, calculated as the ratio of the maximum value of energy that can be obtained from this biomass and the energy of the sunlight that shines on the rice field over the period of a year, is about 0.2%.

The period of growth in the case of rice planted in temperate regions is only from spring to summer, so solar energy cannot be collected all year round. On the other hand, in the tropics the growing season lasts all year. As one example, let's consider how sugar cane is cultivated in Brazil. A mid-summer sun shines all year round, so farmers do not need to cultivate sugar cane just from spring to autumn. Instead, the crop is grown until it is mature, and then it is harvested, irrespective of the time of year. In one region, a continuous growing process lasting for a period of a year and a half is practiced. The average yield for one such region when converted into an annual rate is 50 tons of dry weight per hectare. It is said that if irrigation is used, a yield of 90 tons could be achieved. In that case, the efficiency of biomass production would be slightly less than 1%. We can probably consider this to be the realistic maximum value for production efficiency of biomass on land.

What about the productivity of aquatic plants? Some varieties of green algae, such as chlorella, are known to consume very little of their photosynthesis products themselves. According to one research finding, a biomass production efficiency of close to 5% can be achieved by cultivating chlorella in water under conditions of optimal nutrients and solar irradiation. It is most likely that this value of 5% is the maximum efficiency of biomass production that could be commercialized in the next few decades.

Methods for Solar Thermal Power Generation

Two methods for generating electricity from sunlight that show particular promise for the 21st century are thermal power generation using the same principles as a thermal power plant and direct power generation from sunlight using solar cells.

Solar thermal power generation involves using sunlight to change water into steam and spin a turbine. Several different configurations for doing this are being studied. One example involves heating oil and using it to evaporate steam. As shown in figure 6-2, in the focal point of a concave mirror made of a thin sheet of aluminum, a transparent tube is set through which oil flows and is heated by the focused sunlight. In essence, the sunlight concentrated by the concave mirror is collected to the power plant using the oil. If we define the power generation efficiency as the fraction of the sunlight shining on the concave mirror that is converted into electricity, it is possible to obtain an efficiency of at least 20%. If we can increase the temperature of the oil, the efficiency can be increased even more.

Another method that currently shows promise is a technique that uses a light focusing tower called a "heliostat." In this method, a large number of mirrors are placed in the area around the tower, the reflected light is focused to the collection point in the upper part of the tower, and water is converted to steam for power generation. It is expected that a power generation efficiency of at least 30% can be realized using this method.

The largest drawback of solar thermal power generation is that it only can make use of the direct solar irradiation part of sunlight; it cannot be applied to diffuse sunlight. If the sun is covered by a cloud, the direct solar irradiation is drastically reduced, so in both the focal point of the concave mirror and the collection point of the heliostat tower, reflected light will not be accumulated. Therefore, solar thermal power generation may be an effective technology in deserts where there few clouds to block the direct solar irradiation from the sun, but in highly populated regions that have large energy needs, the number of locations appropriate for this technology are few.

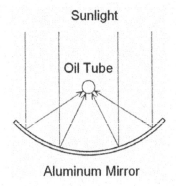

Fig. 6-2: A solar thermal power plant uses a concave mirror to concentrate the sunlight.

Solar Cells Are 40%

Figure 6-3 shows an array of solar cells installed on the roof of a home – for an ordinary home, it is possible to be almost entirely self-sufficient in terms of electricity using this kind of array. The mechanism by which solar cells generate electricity begins when silicon or some other semi-conductor material absorbs sunlight, and the electrons obtain energy. The mechanism up to this point is essentially the same as the first steps of photosynthesis where chloroplasts absorb light. However, in solar cells, these electrons are taken out directly as an electrical current, while in photosynthesis they are used to drive chemical reactions for producing carbohydrates.

We have seen that 95% of the energy of sunlight can theoretically be converted into electricity, so the theoretical maximum efficiency of solar cells is 95%. However, in actual use the efficiency drops considerably. One of the reasons is that efficiency is reduced at the initial steps where light is absorbed by the silicon electrons for exactly the same reason as with photosynthesis. Because there is not just one wavelength of sunlight, it is not possible to use all of the wavelengths optimally. Silicon can absorb electromagnetic radiation with a wavelength of 1 micron or less, which includes a part of infrared radiation, visible light, and ultraviolet radiation. However, most of infrared radiation has a wavelength greater than 1 micron, and that radiation cannot be used. Moreover, in the same way that we saw for photosynthesis, even for photons of highly energetic light, such as violet light, only the amount of energy of a photon of 1 micron infrared radiation can actually be used.

As a result of these factors, an efficiency of more than 40% cannot be achieved using the mechanisms of conventional solar cells (although there are technologies for concentrating sunlight to achieve much higher efficiencies). Furthermore, when we add in other losses due to factors such as impurities in the silicon and inefficiencies in the collection of the electrons, the efficiency of cells that are currently on the market drops to around 15 to 20%. Still, because the electrons that absorbed the light energy can be taken out directly as electrical current, the efficiency of

Fig. 6-3: Solar cells installed on a roof of a house (Courtesy of KYOCERA Solar Corporation)

solar cells is considerably larger than photosynthesis, which involves numerous chemical reaction steps in the production of carbohydrates.

One method for increasing efficiency of solar cells that shows promise for the future is making tandem cells. Rather than just using silicon, tandem cells are made by layering a variety of materials together in order to accommodate a wide range of wavelengths of sunlight optimally. If a solar cell could be manufactured using a continuous range of materials in tandem in such a way that all of the wavelengths of sunlight are perfectly optimized, the ideal efficiency would be 95%.

For example, current silicon solar cells with an efficiency of 15% are made of crystalline silicon. However, simply by layering a thin film of amorphous silicon on the surface, it is possible to raise the efficiency up to about 17%. Currently, the maximum efficiency for solar cells is reported to be 24.4% for silicon, 33.3% when using compound semiconductors, and over 40% for the most advanced concentrating photovoltaic cells.

Does Developing New Technologies Require Too Much Energy?

One of the arguments made by critics of solar cells is the statement that "a large amount of energy is needed to produce solar cells, and it would take 20 years for the cells to recover that energy." However, fortunately, this argument is incorrect.

The idea of using an energy system, such as solar cells, to save energy is based on the assumption that the amount of energy produced by the system will replace the consumption of an amount of conventional energy resources such as fossil fuels that is significantly larger than the amount of energy resources required to make the system in the first place. The length of time that an energy system must be operated to recover the energy consumed during the manufacture of the system is called the "energy payback time." Here we will take a look at what the energy payback time for solar cells is.

Solar cells are constructed from a variety of materials. The strength of the cell is provided by a frame of aluminum, the surface of the cell is protected by glass, and the power generating part of the cell is made of a semi-conductor such as silicon. Furthermore, in order to adjust for the imbalance of power generation between night and day, or between rain and shine, some kind of mechanism for storing the generated electricity or for exchanging power with the local electric power company is also needed. However, after listing up and evaluating all of the materials and processes that go into making solar cells, from the mining of resources to the manufacture of the silicon thin-film and the assembly of the whole cell, a study by the Society for Chemical Engineering of Japan found that in the case where cells manufactured using current technology are installed on rooftops in Japan, the energy payback time is only about two years.

Comparing Solar Cells and Biomass

Among the technologies for using renewable energy, solar cells and biomass are a pair of technologies that show great promise as sources of renewable energy for the future. They have considerably different characteristics. In terms of energy efficiency, solar cells are superior to biomass. We have seen that carrying out the cultivation of sugar cane in Brazil under the optimal conditions of sunlight and irrigation results in an efficiency of 1%. If we estimate that the silicon solar cells on the market will be able to reach an efficiency of 20%, the difference is twenty-fold. This means that in order to obtain the same amount of energy, one twentieth of the area is sufficient if we use solar cells.

On the other hand, from the viewpoint of energy payback time, biomass has the upper hand. Figure 6-4 shows a concept diagram for a system where eucalyptus trees are planted and used as biomass in Western Australia. A circular area of land 25 km in diameter is divided into 12 sections like a clock. A drying site and power plant are set up in the center. Alternatively, in place of the power plant a chemical plant for manufacturing methanol or fuel oil could be used. Of the 12 sections, 11 sections are kept planted, and each year one section is harvested for biomass that is collected to the drying site at the center. At this scale, the system can produce an amount of fuel oil each year that is equivalent to 150,000 tons of crude oil, or if the system is used to generate electricity, it will have a generating capacity of 100,000 kW, which is the equivalent of a mid-sized coal-fired power plant. This system has been designed and evaluated based on the assumptions that the planting,

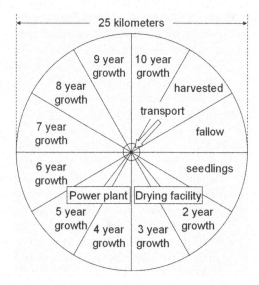

Fig. 6-4: A biomass utilization system

cultivation, and harvesting are all done mechanically and that an appropriate amount of fertilizer is applied. According to that evaluation, regardless of the form in which the final energy is obtained, the energy payback time is in the range of just 5 to 75 days. Therefore, the energy investment for this biomass system can be recovered in a much shorter time interval than in the case where solar cells are installed on rooftops, which we saw would take two years. Because the initial investment of biomass production can be recovered so quickly, biomass is probably better suited for quick applications than solar cells.

The fundamental differences between biomass as less efficient but more easily implemented and solar cells as more efficient but more costly and difficult to start up suggests an approach where biomass technology is used to facilitate the transition to solar cells. Land that is secured for cultivation of biomass and conversion to fuel could gradually be replaced with solar cells, which could increase the energy generation rate per unit area by more than twenty times.

3 Hydropower and Wind Power

Potentials Large and Small

In general, assessments of renewable energy resources vary greatly depending on how the assessment is made. For instance, an upper limit for the quantity of hydropower and wind power resources can be estimated from the energy balance at a global scale, and a lower limit giving the amount of resources that we know for certain to exist can be obtained by adding up the results of individual surveys made at each resource site. However, the difference between these two values is large.

The size of a water resource for hydropower generation is its potential energy, which is just the amount of water multiplied by its height. The average rainfall around the world is about 1 meter per year. If we consider that this rain on average falls from a height of 1000 meters, then the amount of resources for hydropower generation if all of the rain water were collected at this height would be more than double the current global amount of energy usage.

However, to recover this amount of hydropower resources would require doing something drastic like collecting rain in a plastic sheet stretched over the entire sky of the earth and dropping that water through a 1000 meter long turbine to generate electricity. If we take just the part that falls to dry land, the value becomes one fourth, which is about half of the worldwide amount of energy usage. On the other hand, adding up the results of surveys of flow rates and heights of all known rivers around the world, the total amount of undeveloped hydropower resources remaining appears to be approximately the same order as the current developed hydropower resources (one estimate gives the potential hydropower resources that are economically viable as 9400 TWh, which is four times the current developed hydropower resources), which produce 5% of the total amount of energy use. Therefore, there is a more than two-fold difference between the estimate of the potential energy in the rain that falls to dry land and the estimate of water resources from surveys. It is difficult to imagine that any large rivers still remain undiscovered on the earth's surface, so we should probably take the survey-based estimate of unused water resources as the basis for decisions regarding hydropower.

If we include the westerlies and other major winds that blow at high altitudes, natural resources for wind power have an energetic value that is greater than that of water resources. However, when we limit the altitude of the wind resources that we can use, the amount becomes much smaller. For the height of current wind power generation facilities, the amount of wind resources is about the same level as the worldwide amount of energy consumption. Even this is a considerable amount of resources. However, with the wind powered electric generation technology available today, the generator will not work under conditions of weak winds, and when a wind blows that exceeds the design strength, then the operator must shut the generator down. As a result, the generator typically operates for only about 70 to 80 percent of the time on average, and even when it is operating, much of the time it is not operating at its maximum power output. In fact, a wind turbine that is rated at 1000 kW will typically only produce about 20% of its maximum power output each year. Furthermore, when we consider all of the conditions that are necessary for current wind powered electric generation, such as having a stable wind, having a low local population, and not being too far from a region with a demand for energy, it is not clear how many appropriate sites are in existence. Data with the reliability of the survey results for hydropower have yet to be obtained.

The Natural Circulations Are Concentrated

Hydropower is an excellent renewable energy that is clean and can be transformed with almost 100% efficiency into electricity, as we saw in Chapter 2. These benefits come from using water that is collected over a wide area over a relatively long period of time. Therefore, the major problems of solar energy that we saw earlier in this chapter, which are low density and rapid temporal fluctuations, are solved through the circulation of water. Although wind cannot be collected behind a dam, it also benefits from the circulation of air, which can collect the kinetic energy of wind over a wide area and direct it towards the position of the wind turbine.

However, one important problem with hydropower development is that valuable land becomes submerged. Take the example of the "three gorges dam" in China. This is a huge dam, whose construction began in 1994 and is scheduled to be completed in 2011. The completed dam will have a generation capacity of 22,500,000 kW, which is more than 2% of the total power generation capacity in China. It is said that 660 km² of land was submerged as a result of construction of this dam and that 1,130,000 people were forced to move. One way to alleviate this problem is to make a large number of small dams as shown in figure 6-5. Remember that hydropower gets electricity from the potential energy of water, which is determined by the product of the water amount and height. Therefore, so as long as we

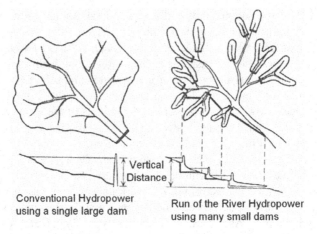

Conventional Hydropower
using a single large dam

Run of the River Hydropower
using many small dams

Fig. 6-5: Conventional hydropower versus run of the river hydropower
Note: the shaded area shows land that is flooded by the dam.

accumulate the same amount of water over the same vertical distance by building
many small dams in the catchment area flowing into the location where the single
large-scale dam was to be constructed, we can generate the same amount of electricity, even though the total land area flooded is much smaller. This way is also
easier to implement economically.

4 Geothermal Energy and Tides

We can imagine geothermal power generation as digging a deep hole and burying
a U-shaped steel pipe that reaches the hot mantle of the earth. When we pump water
into the pipe, it will turn to steam as it travels down to the earth's mantle and back,
and that steam can be used to turn a turbine and generate electricity. Currently,
there are still only a small number of applications of this technology, mainly
because only heat close to the surface can be used economically. Places that have
geothermal heat near the surface are places where hot springs and geysers most
easily upwell. Such locations are often natural parks or tourist attractions where
development is difficult, so it is not likely that the use of geothermal energy will
expand rapidly.
   On the other hand, the amount of heat contained within the earth is tremendous,
and if we could find a way to tap into that energy, the amount of geothermal
resources would rival the energy from the sun. Many ideas for geothermal technologies have been suggested, such as power generation using high-temperature rocks
and ways to tap in to geothermal resources deep below the earth's surface in a cost
effective manner. However, methods for actually implementing these ideas have
not yet been established. One example of a concrete method that has been proposed
for power generation using high-temperature rocks involves jetting water out of a

steel pipe underground at extreme pressures. The high pressure water jet breaks up the high-temperature rocks underground and is heated through contact with the rock fragments. The water is then collected at a high pressure and high temperature at a different location on the surface, where it is used to generate electricity at a thermal power plant. In order to extract heat from large rocks, they must be broken up into small enough pieces that the contact area between the water and the rock fragments is sufficient. Experiments are being conducted, and promising results have been reported. However, the technology development is still at the stage of feasibility research studies, and it has not yet reached a level where one could say that the prospects are sufficiently developed for practical application.

The ebb and flow of the tides caused by the gravitational attraction of the moon and the sun can be used to obtain energy. All we need to do is build a flood gate at the entrance of a bay. We open the flood gate when the tide is rising, and allow the tide to flow into the bay. Then when the tide begins to ebb, we close the flood gate and force the water accumulated inside the bay to return to the ocean through the same kind of generator that is used in hydropower plants.

One power plant that uses this kind of tidal electricity generation has actually been in operation since 1967 in Reims, France. The power generation capacity of the plant is 260 kW, which is about the size of a small hydropower plant. However, due to the large number of features that are required for the coastal region to be suitable for this kind of electricity generation, such as large tides and bays with small mouths, finding appropriate sites is difficult. Therefore, we probably cannot expect a large contribution from this technology.

In this chapter, we have seen that energy from the sun and the earth's core exists in practically limitless amounts, and its quality is also high. However, the energy from the sun is difficult to harness because of its low density and temporal instability, and few reliable methods for harnessing energy from the earth's core have been sufficiently developed for commercial applications. Probably the only methods that could reliably overcome the problems of density and instability and could be deployable on a large scale within the 21st century are solar cells, solar thermal power generation, biomass, and deep geothermal power generation. While the technologies currently available are still expensive and difficult to commercialize, it is almost certain that some excellent technology that is well suited for a material-recycling society could be developed in the not too distant future. However, in order to realize that possibility, we must invest our efforts in research and development of the most promising technology options existing today.

Development of technologies and systems that can generate large amounts of energy at the quality and cost of what is currently obtained from fossil fuel resources will take some time. The task of replacing the majority of fossil fuel resources with renewable energy will most likely take until the end of the 21st century. Oil, and possibly all of the fossil fuels that we currently depend on, will be completely depleted on this time scale. Consequently, together with speeding the development of renewable energy, we must work to reduce fossil fuel consumption in order to make time for the expansion of the practical application and scale of technologies that use renewable energy.

Chapter 7
How to Make a Sustainable Earth

In this chapter, we will summarize the ideas that we have introduced in the previous chapters and use them to develop "Vision 2050," a roadmap for achieving a sustainable human existence on the earth. In developing "Vision 2050," we will take a critical look at what the requirements will be for human society in 2050. Those requirements will give us the infrastructure necessary to support all humans on the earth in 2050. We will then see how we can achieve this necessary infrastructure through technology and well-coordinated development in both developed and developing countries.

To give a quantitative description of "Vision 2050," we will need to choose a base year for our discussion. We have chosen the year 1995 as the base year for "Vision 2050." We chose this year based on the availability of data as well as the milestone event that occurred in the late 1990's – the birth of the six billionth person on the planet. The first year of the millennium (or the last year of the previous millennium) may have been a more memorable choice. However, much of the dialog on attainment of a sustainable earth has centered on the Kyoto Protocol for CO_2 emissions reductions. The Kyoto Protocol, which we will look at next, takes 1990 as the base year. We have split the difference and used the year 1995.

1 The Significance of the Kyoto Protocol

The Inevitability of Global Warming

In December 1997, COP3 (the third session of the Conference of Parties to the United Nations Framework Convention on Climate Change) was held in Kyoto, Japan, and the Kyoto Protocol was adopted as an outline for reducing the emissions of CO_2, focusing in particular on the developed countries. The gist of the protocol was that, relative to 1990 levels, by 2010 Europe, the U.S. and Japan would reduce their emissions of CO_2 by 8%, 7% and 6% respectively.

Hiroshi Komiyama and Steven Kraines
Vision 2050: Roadmap for a Sustainable Earth.
© Springer 2008

Many experts have expressed opposition to the Kyoto Protocol, claiming that "the reduction targets are too small," "flexible measures such as emissions trading will undermine the actual effect on emissions reduction," or "it will cause an adverse impact on global economic growth." So how valid are these criticisms? We have seen in Chapter 1 that the phenomenon of global warming is real, and that even if we could reduce emissions rates to those of 1990, global warming is likely to cause serious problems by the middle of the 21st century. Reducing CO_2 emissions in Europe, the U.S. and Japan by a small percentage is hardly enough to prevent the looming problems of global warming. The effectiveness of emissions trading also raises serious questions. Finally, depending on what mechanisms are used to implement the Kyoto Protocol, we cannot be sure that it will not adversely impact global economic growth.

However, there is no excuse for doing nothing. If measures for reducing CO_2 emissions are implemented in accord with the principles advocated in this book, the emissions reduction goals stipulated by the Kyoto Protocol could be achieved with at most only a small negative impact on economic growth. As a basic rule of thumb, we can consider reducing CO_2 emissions in the short term to mean reducing the use of energy. We saw in the previous chapters that there is still considerable potential for reducing energy use in both "making things" and "daily life." Moreover, in the long term these reductions will save money as well in both the manufacturing sector and the private sector.

In addition to making a small but concrete contribution towards mitigation of global warming, the Kyoto Protocol is a powerful symbol. Until now, human activity has traced a path focused only on expansion, and in response energy consumption has increased steadily. Thus the Kyoto Protocol is a milestone, marking a consensus among nations including the U.S. that we must make some changes to this headlong pace of expansion.

However, as we saw in figure 1-1, even after the Kyoto Protocol was agreed upon, the concentration of CO_2 in the atmosphere has continued to rise. It seems unlikely that even the moderate reduction stipulated by the agreement will be met by the deadline of 2010. If we continue in this way, we must face the possibility that a tremendous increase in global warming during the 21st century is inevitable.

A Gap Between the Developed World and the Developing World

The success of the Kyoto Protocol depended the stances taken by the U.S., the world's largest consumer of energy at the time, and by developing countries, whose demands for energy are predicted to constitute the bulk of increased energy consumption in the future. The U.S., which consumes one fourth of the world's energy, has made low energy prices a national strategy. The price of gasoline in the U.S. during the 1990's was about 30 cents per liter, and the price of electricity for industrial use was about 4 cents per kilowatt-hour. For comparison, in Japan, Korea and most of OECD Europe gasoline cost almost one dollar per liter, and in

Japan electricity cost more than 10 cents per kilowatt-hour. Through these low energy prices, the U.S. subsidized manufacturing and encouraged the use of automobiles. However, by bringing the U.S. into the discussion of how to reduce CO_2 emissions, participants in the Kyoto Conference, including Japan and Europe, hoped to pressure the U.S. into making reductions. Unfortunately, even though the White House was environment-friendly, U.S. leaders were not confident that the American people would support reducing energy consumption. Partly to avoid facing a heavy domestic backlash, the U.S. made their participation in the Kyoto Accords conditional on the inclusion of developing countries, where most of the future increase in CO_2 emissions is predicted to occur.

But the argument put forth by the developing countries was irrefutable. Of the total global CO_2 emissions, 75% are from the developed countries while the developing nations, home to 75% of the world's population, produce only 25% of the total CO_2 emissions. Clearly, the developing countries cannot be expected to take responsibility for current CO_2 emissions. Moreover, to increase their standard of living, developing nations must increase their consumption of energy in the future. Although this increased energy consumption will be accompanied by an unavoidable increase in CO_2 emissions, developing nations cannot be forced to maintain a standard of living below that of the developed world. And the inevitable increase in CO_2 emissions becomes even clearer when we consider the importance of continued economic growth in developing countries to the economies of the rest of the world.

According to the U.S. Department of Commerce website, on July 19, 1999, the human population of the earth reached 6 billion. As of the beginning of 2008, the population has become 6.6 billion. By 2025, the population is predicted to be almost 8 billion, and by 2050, about 9 billion. In Japan, Europe, and most of the other developed countries, national populations have peaked or are nearing their peaks. Therefore, most of the increase in the world population – an increase of 3 billion by 2050 – will occur in the developing countries.

As noted in Chapter 2, the current global population of more than 6.5 billion people consumes 7.5 billion tons of fossil fuel resources per year. Therefore, the global average fossil fuel consumption is slightly more than one ton per person. In comparison, the average amount of fossil fuels used per person in Japan, England, and Germany is about 2.7. In the U.S. the amount per person is over 5.5 – more than double the average of other developed countries. The average for Japan and the OECD countries of the EU is about 2.4, a value that is representative of developed countries other than the U.S. So if we assume that all 7.5 billion inhabitants predicted to be living in developing countries by 2050 will consume fossil fuels at this rate, the resulting fossil fuel consumption would be about 18 billion tons per year in the developing countries alone. Even if we assume that the consumption rate of 4.5 billion tons per year in the developed countries does not increase at all, the total global annual fossil fuel consumption rate in 2050 would be nearly 23 billion tons. This rate is almost four times the current rate of fossil fuel use, and about three times the total annual energy use today, including hydropower and nuclear power.

We have seen that the ratio of confirmed oil reserves to the current annual consumption rate is 40 years. If our consumption of oil grows by three or four times this rate, by 2050 almost all known reserves will be depleted.

So how about the other fossil fuels?

Hope, but Do Not Expect Too Much . . .

There are many opinions about the lifetime of energy resources. Coal is said to have about 150 years of reserves as of 2007, so some experts claim that there is no need for concern. However, the prediction of 150 years is calculated based on the current rate of coal consumption, which is now much lower than that of oil. If we assume that coal will replace oil when oil is depleted, the lifetime of coal reserves will, of course, become shorter. For example, if the four-fold increase in energy consumption that we have calculated here is covered entirely by coal, coal will be depleted in just a couple decades. And most important, we must not forget that CO_2 emissions from coal are 1.5 times greater per unit of energy than emissions from oil.

Many people have put their faith in natural gas as a replacement for oil. The main component of natural gas is methane. Methane hydrides – ice-like substances formed from mixing water and methane – are said to exist in large quantities in the ocean floors and in the frozen soil of Siberia. Although many deposits of methane hydrides have indeed been confirmed, there have been few studies on how much energy would be consumed in extracting and processing this substance into usable energy. However, there is no doubt that if we were to use methane extracted from an ice-like substance on the ocean floor, it would consume more energy than is currently used in mining coal or in retrieving oil and gas from offshore oil fields. In addition, methane is also one of the greenhouse gases targeted by the Kyoto Protocol, and, per unit mass, the greenhouse effects of methane are over 20 times larger than those of CO_2. There is a concern that methane could be released into the atmosphere when methane hydrides are extracted, contributing further to the global warming effect.

Finally, there are unverified claims that a form of methane exists which is not the product of fossilization. The claim is that deep underground, inexhaustible pockets of methane exist that were produced directly from water and CO_2 long ago. It has been shown in laboratory experiments that if water and CO_2 coexist in the presence of some metal such as iron, then – under conditions of great heat and pressure – methane can form. So it is possible that these reservoirs of methane exist. However, there is as yet no proof of such reservoirs, nor have any been discovered in the several experimental drillings that have been carried out. It would be foolish to gamble the future of the human race on the chance that this theory will pan out.

We must assume that not just oil but all fossil fuel resources will be scarce by around 2050. And we must honor the agreement made in Kyoto, not only because it is an international agreement but also because it is a necessary first step towards planning the further reduction of CO_2 emissions and fossil fuel consumption into the future. Indications of global warming, oil depletion, and massive of waste are already apparent. We cannot deny the possibility that we are heading towards a potential catastrophe in the middle of the 21st century.

2 Vision 2050: A New Road to a Sustainable Earth

Three Preconditions

Okay, let's try to find a road out of this catastrophic situation. We will call this road "Vision 2050." But first we must set a few preconditions for our journey.

The first precondition is that developing countries must be guaranteed the right to modernize. No one in the developed world could convincingly argue that the citizens of developing countries should maintain their current living standards. While some might argue that people in developing countries are being seduced into adopting a modern civilization that consumes large amounts of energy, this argument is hardly persuasive when put forth by those enjoying a life of luxury to consign others to a life of poverty.

The second precondition is that the energy conservation required to achieve Vision 2050 cannot be based on unrealistic expectations of people making radical shifts in their lifestyles. The energy conservation needed to achieve Vision 2050 can be divided roughly into energy savings from changes in lifestyle and savings from increased efficiency through improved technologies. In Chapters 3 and 4, we have looked at potentials for savings through improved technology. However, it is more common for a discussion of energy conservation to begin by recommending changes in lifestyle. Although the primary goal of this book has been to show the potential for technologies to help us to achieve a sustainable earth, let's now consider briefly the potential savings from changes in lifestyle.

Many people today feel that there is something wrong with the societies that have developed in the last century – societies that encourage consumption. Is it really necessary to blast the air conditioner in the summer? Is it really sensible for stores to give us so many plastic bags, which we eventually throw away? Many people feel in their hearts that major lifestyle changes are necessary. And energy savings through lifestyle changes would, of course, help reduce energy consumption. For example, a 10% savings of energy through lifestyle changes would reduce energy consumption by 10% and thereby reduce the use of fossil fuels and CO_2 emissions by approximately 10%.

Another important lifestyle change would be to cut down on waste. We should be able to establish agreements among manufacturers, distributors, retailers, and consumers to cut back on excessive packaging and wasteful copying. If we are committed to conserving energy, we might begin using both sides of paper. We might prohibit driving cars for personal use in city centers. Such strategies for reducing energy consumption are within the realm of possibility, and in Vision 2050 we assume that there will be a contribution to savings from these lifestyle changes. However, it is dangerous to rely too much on the effect of these changes. We have seen that the increased energy consumption that will occur in the developing world may exceed three times the current energy use. It is unrealistic to expect that sustainability can be achieved through energy savings alone. We need to complement efforts to save energy through lifestyle changes with ways to increase the efficiency of energy consumption in both "making things" and "daily life" through technology.

The third precondition is that, as we saw in Chapter 6, the likelihood that we will succeed in replacing fossil fuels with renewable energy by 2050 is, unfortunately, almost zero. Many people have high expectations for renewable energy. However, aside from hydropower and the use of wood for fuel in developing countries, the contribution of renewable energy to total energy today is 1% – not nearly enough to form the basis for large-scale dependence on renewable energy by 2050. The problem is that it is difficult to engineer a system that can transform an energy source that is thinly spread out and variable over time into convenient forms of energy such as electricity and vehicle fuels that can be used whenever we want. So we must face the fact by 2050, we will still be somewhat dependent on fossil fuels.

The Basic Concepts

Figure 7-1 shows the levels of energy use for several scenarios. The situation in the base year, 1995, is shown as scenario (a). In 1995, the equivalent in carbon units of about 7.5 billion tons of energy resources was consumed. This includes 6 billion tons of fossil fuels plus 1.5 billion tons of non-fossil fuel energy sources, mainly wood, hydropower, and nuclear power. In the top figure for scenario (a), the dark part represents the 6 billion tons of fossil fuels, and the light part shows the contribution from the non-fossil fuel sources.

We saw earlier that the 75% of the world's population living in developing countries, 4.5 billion people, consume just 25% of the total fossil fuel energy resources: 1.5 billion tons. As a rough estimate, we will consider that half of the total non-fossil fuel energy, about 0.75 billion tons carbon equivalent, is used in the developing countries (mainly biomass and hydropower) and the other half is used in the developed countries (mainly hydropower and nuclear power). Therefore, the 1.5 billion people in the developed world consume about 5.25 billion tons of energy resources and the 4.5 billion people in the developing world consume about 2.25 billion tons of energy resources. This results in an average energy use per person of 3.5 in developed countries and 0.4 in developing countries. The average use of fossil fuels per person is 3.0 in developed countries and 0.3 in developing countries. In the bottom figure for scenario (a), the hatched part

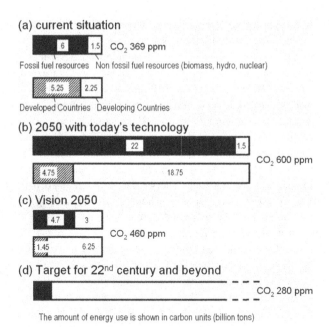

(a) current situation

6   1.5   CO$_2$ 369 ppm

Fossil fuel resources   Non fossil fuel resources (biomass, hydro, nuclear)

5.25   2.25

Developed Countries   Developing Countries

(b) 2050 with today's technology

22   1.5

CO$_2$ 600 ppm

4.75   18.75

(c) Vision 2050

4.7   3

CO$_2$ 460 ppm

1.45   6.25

(d) Target for 22nd century and beyond

CO$_2$ 280 ppm

The amount of energy use is shown in carbon units (billion tons)

Fig. 7-1: Energy scenarios and CO$_2$ concentrations
Note: for each scenario, the top figure shows the distribution of energy consumption between fossil and non-fossil energy resources, and the bottom figure shows the distribution of energy consumption between developed and developing countries.

represents the 5.25 billion tons of energy resources in carbon units used by developed countries, and the light part shows the 2.25 billion tons of energy resources used by developing countries.

If in 2050, the 7.5 billion people predicted to be living in the developing countries have reached energy consumption rates equal to those of the developed countries today (excluding the U.S.), then we have seen that about 18 billion tons of fossil fuels will be necessary to meet the demands of those countries. We will assume that the amount of non-fossil fuel energy used in 2050 will be the same as it is today. As a result, the energy use per person in the developing countries will be about 2.5, which is considerably less than the current average for developed countries of 3.3. If the energy consumption of the developed countries remains the same as it was in 1995 – the equivalent of 5.25 billion tons of fossil fuels – and if the demand for energy in developing countries rises to 18 billion tons of fossil fuels plus the 0.75 billion tons of non-fossil fuel energy used today, then the total consumption of energy per year on the planet will be 24 billion tons of fossil fuel equivalent. Even if the people in the developed world were to reduce their fossil fuel consumption from the current average of 3 tons per person to the OECD Europe average of 2.4 tons per person through intensive energy savings efforts, they would still consume about 4 billion tons of fossil fuels, giving a total fossil fuel consumption of 22 billion tons per year and a total energy consumption of about 23.5 billion tons per year. This is over three times the amount of energy used today and is represented in the figure as scenario (b).

We have seen that when we use energy for some purpose or function, the energy efficiency differs remarkably depending on the technology. For example, driving a car for a distance of 10 km requires a different amount of energy depending on whether the car is powered by a normal combustion engine, a hybrid engine, or a fuel cell engine. If efficiency is increased, the same function of driving 10 km can be performed with that much less energy. The 23.5 billion tons in scenario (b) is the projected energy consumption in 2050 based on today's technologies and social institutions. If we can significantly increase energy efficiency, we can perform the same functions with less energy. Even if our need for energy-based functions triples by 2050, if the energy efficiency in performing these functions also triples, we can sustain the increased demand for the function while keeping energy consumption at the 1995 level.

However, even if we could keep the amount of energy consumption worldwide at the level in 1995, if we continue to rely on fossil fuels as the source of that energy, the problems of global warming and the depletion of fossil fuel reserves will remain unsolved. To address these problems, we need to bring into play as much renewable energy as possible by 2050. If we could develop an amount of renewable energy equal to the total amount of non-fossil fuel energy used today, about 1.5 billion tons carbon equivalent, then the amount of fossil fuel consumed each year could be reduced to 4.7 billion tons, which is just a little more than three quarters what it is was in 1995.

Scenario (c) in figure 7-1 shows the basic concept of Vision 2050. First, although the total energy-related functions required in the world will increase to three times that of the base year of 1995 shown in scenario (a), mainly due to the modernization in developing countries whose total population will increase from 4.5 to 7.5 billion, we will triple the efficiency of energy consumption for meeting this requirement. As a result, the actual energy consumed per person will be less than 1 ton carbon equivalent per person in both developed and developing countries, and the total energy consumption will remain almost the same as it is today. Second, by introducing an amount of renewable energy equivalent to the total amount of non-fossil fuel energy currently produced, the use of fossil fuels will be reduced to almost three quarters of what it was in 1995.

Scenario (d) depicts a situation for the 22nd century where only a tiny amount of fossil fuels is used together with far more renewable energy than is shown even in scenario (c). By following the road that is laid out in Vision 2050, we can make this scenario a reality by continuing to reduce fossil fuel consumption and to increase the use of renewable energy through the second half of the 21st century.

Figure 7-2 shows another way of looking at the three main scenarios in figure 7-1. In Chapter 1, we introduced an equation for the sustainability of human existence on the earth, where the impact of humans on the earth equals the product of the human population, the affluence of that population as measured by the functions of products and services consumed per person, and the impact on the earth of providing one unit of function, such as the energy resources consumed.

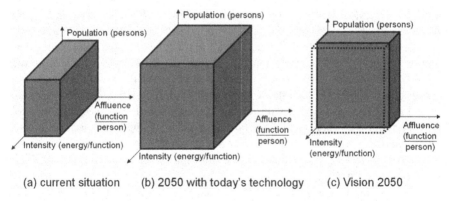

(a) current situation   (b) 2050 with today's technology   (c) Vision 2050

Fig. 7-2: Human impact on the earth for three scenarios

In figure 7-2, we show scenarios (a), (b), and (c) as three dimensional boxes whose volumes represent the impact of human civilization on the earth. In scenario (a), the population is lower and the affluence is smaller, mainly because of the low standard of living in developing countries. But the intensity, which is the inverse of energy efficiency, is high, so the overall impact on the earth is fairly large. In scenarios (b) and (c), population has increased about 50%, and affluence has almost doubled. The difference is that while the intensity in scenario (b) is the same as in scenario (a), it is one third in scenario (c). In fact, the volume of the box extending out to the dotted line in scenario (c) is almost the same as the volume of the box in scenario (a). Furthermore, when we consider the introduction of non-fossil fuel energy resources under Vision 2050, the actual impact on the earth in terms of fossil fuel resource consumption is just the volume of the grey box in scenario (c). This is another way of looking at Vision 2050.

A Crossroads

The increase in the concentration of CO_2 in the atmosphere is approximately proportional to the rate of emission of CO_2 by human activity. And currently, the concentration of CO_2 in the atmosphere is increasing at an annual rate of 2 ppm. So if we continue to emit CO_2 at the current rate, in fifty years – even without considering the population growth and economic growth in developing countries – the increase would be more than 100 ppm. Therefore, by 2050, the concentration of CO_2 in the atmosphere will rise from the 1995 value of 369 ppm to more than 469 ppm.

Let's use this approximation to estimate the CO_2 concentration in scenario (b) from figure 7-1. If we assume that the annual rate of fossil fuel consumption will increase linearly from 6 billion tons in 1995 to about 22 billion tons in 2050, a simple calculation shows that the concentration will reach about 600 ppm. This value far exceeds a doubling of pre-industrial levels. On the other hand, in the case of scenario (c) – where the rate of consumption of fossil fuels in 2050 is three quarters what it was in 1995 – the concentration will be 460 ppm. While considerably less than the 600 ppm of scenario (b), this is still a huge increase from the value of 369 ppm in 1995. Must we really accept this as the lowest level that we can hope to achieve? In fact, this value is only slightly less than the 469 ppm that would result if we continued with the current situation. It may seem like we will have done little to improve the situation. Remember, though, that per capita consumption of fossil fuels in 2050 will be reduced to 75% of the rate in 1995, so the rate at which the CO_2 concentration in the atmosphere increases after 2050 will be reduced proportionally.

At that point, if we can move to scenario (d), we will be able to slow the increase of CO_2 concentration even further, and eventually it will begin to decrease as CO_2 in the atmosphere is absorbed by the ocean. Therefore, although it is probably impossible to completely avoid global warming from the increase of CO_2 concentration in the atmosphere, if we can achieve Vision 2050, we will have paved the way for reducing CO_2 emissions in the future thereby reaching a stable atmospheric CO_2 concentration and an end to increased global warming by the 22nd century.

Obviously, an important factor in the future of the earth is the increase in the human population. However, as income levels in developing countries increase to match those in developed nations, population growth is predicted to decelerate. This relationship between income level and population growth has been confirmed by experience. So if, by 2050, the 7.5 billion people living in developing countries reach a standard of living comparable to that in developed countries today, the world's population should start to decline.

When our descendents look back on the history of this century, they will surely see the year 2050 as a milestone. Will a lifestyle of mass production and mass consumption spread to developing countries, causing energy consumption to exceed three times that of today? Will waste materials cover the surface of the earth? Will the concentration of CO_2 in the atmosphere increase to more than double its pre-industrial value? Or will we – through recycling our waste materials, tripling our energy efficiency, and doubling our use of renewable energy together with making moderate changes to our lifestyle – be successful in creating a path to a sustainable human community by the 22nd century? The crossroads that lies before us will determine upon which road this milestone will be laid.

3 Making Vision 2050 a Reality

Vision 2050 has three main parts: a three-fold increase in energy efficiency, a two-fold increase in use of renewable energy, and conversion to a system of material recycling. Now let's see how it will be possible to meet these conditions by 2050.

(1) A Three-fold Increase in Energy Use Efficiency

Reduce Energy Used in Transport, Homes and Offices to One Fourth

First, we can reduce gasoline used by cars to one-fourth what it was in 1995. We have already seen that we can cut energy consumption 75% by reducing a vehicle's weight and using hybrid engines, so doing that would be enough. In fact, as of 2007, new hybrid vehicles on the road have already cut energy consumption by about 50% compared to automobiles in 1995. Alternatively, we could combine these technologies with ways to reduce friction such as designing new kinds of tires. Or perhaps we could use fuel cells as a power source. What ever combination we use, reducing energy consumption for passenger cars to one fourth by 2050 should be an achievable target. And the same improvements in efficiency can be achieved for other vehicles, such as buses and trucks. If we take the average life of vehicles to be ten years, by 2050 the fourth generation of automobiles will be rolling off the production line. Consequently, it should be well within the realm of possibility to convert just about all of the vehicles in operation to this level of fuel efficiency by 2050.

We can effect a similar improvement in the energy efficiency of homes and offices. The main form of energy consumed here is electrical. Looking back to the data that we discussed on the use of energy in Japan, even if we consider that the average efficiency for thermal power plants in Japan today is 43% (using the high heating value), still fully two thirds of total energy resources consumed in Japan through "daily life" activities in offices and homes is used as electricity. Furthermore, the fraction of total energy consumed as electricity is increasing each year, so we can estimate that by 2050 around 80% of the total energy resources used in homes and offices will be used as electricity. Therefore, when we look at the possibilities for energy conservation in "daily life" activities at homes and offices, it will be reasonable to assume that all of this energy comes from electricity.

We could triple the efficiency of air conditioners and other heat pumps by increasing the efficiency of compressors and decreasing the temperature difference in heat transfer. With additional measures such as increasing insulation in houses, we could increase the overall efficiency of heating and cooling by five times. Refrigerators are also heat pumps. Although some loss of efficiency, such as that from opening and closing the refrigerator, is unavoidable, we should be able to increase their efficiencies as well. In fact, during the period from 1995 to 2005, through advances in vacuum insulation and technologies for reducing energy loss when opening the refrigerator by using sensors and compartmenting the space with multiple doors, energy efficiency of refrigerators has tripled already. For lighting, we could develop light-emitting devices with twice the efficiency of fluorescent light bulbs. Then by reducing the proportion of highly wasteful incandescent bulbs, we could triple the efficiency of lighting homes and offices. Although the size of televisions will probably continue to increase, through the use of low-energy technologies such as LCD displays and semi-conductors, we could double the efficiency of televisions. Energy conservation for other appliances such as vacuum cleaners, rice cookers, and microwave ovens may be more difficult, but because these are in use for relatively short periods of time, the total energy they consume is not so large.

If all these improvements in efficiency were effected in homes and offices, we could reasonably expect to reduce energy consumption by up to 60% of 1995 levels.

Working from the other side, we can reduce the amount of fossil fuel consumed per unit of electricity that is used by these devices by improving the efficiency of power plants in generating electricity. In Vision 2050, we will set our goal to reduce fossil fuel consumption in this way by one third. We could achieve this by increasing our efficiency in generating electricity from the 1995 level of 38% to a level of 57% in 2050. Although the lifespan of electric power plants is long, we can assume that by 2050, all but the newest plants will have been replaced. Already, combined cycle power plants exist with efficiencies of 53%. If the top power plants in 2050 achieve efficiencies of 65% and if the most advanced power plants existing today with efficiencies of around 50% to 53% are the oldest plants remaining in 2050, that will raise the average efficiency to 57%. Note that as in the previous chapters, these thermal power plant efficiencies are all in terms of the higher-heating values.

Another possibility for increasing efficiency is that distributed electric power systems will become widespread. For example, by 2050, fuel cells may be available with a conversion efficiency of fuel to electricity of about 50%. Because fuel cells also generate usable heat, they can be used for co-generation of heat and power in individual buildings. Alternatively, other technologies for generating electricity on a small scale, such as combinations of small-scale gas turbines and steam turbines, might be developed to create highly efficient co-generation systems. When the value of the useful heat is converted to electricity and added to the total system output, it might be possible using such co-generative systems to achieve an overall efficiency equivalent to an electric power generation efficiency of 57%.

If we combine the effects of reducing energy consumption by 60% (through increased efficiency of appliances) with the effects of reducing fossil fuel consumption by 33% (through increased efficiency in generating electricity), we see that the consumption of fossil fuels for electricity supplied to homes and office buildings can indeed be reduced to $(1 - 0.6) \times (1 - 0.33)$, or about 25% of today's consumption rate.

Reduce Energy for Material Production to One Third

We can reduce the energy consumed in producing materials, particularly metals, through a combination of recycling, developing new technologies, and transferring technology. First, we can cut energy consumption by expanding the recycling of the different kinds of materials we use. If the current rate of producing goods from natural resources were to continue unabated, by 2050 we would reach the point where future production of all of the most important basic materials could be carried out through the use of scrap. However, in fact the proportion of products made from natural resources will decrease as the accumulation of human artifacts increases and recycling is expanded. Therefore, we probably will not reach the point of complete saturation by 2050.

Let's suppose that by 2050 scrap will constitute 80% of the material used in creating new products. By producing 80% of iron from recycled metal instead of iron ore and by melting the recycled metal in furnaces heated by fossil fuel instead of electricity, we could reduce energy consumption per unit of iron produced to one third that in 1995. Even now, aluminum can be produced from recycled materials using only one tenth the energy required in production from natural bauxite. So even if the efficiency of aluminum recycling does not improve at all, at the point where 80% of aluminum is recycled, the total energy consumed in production will decrease to about one fourth what it was in 1995.

Under Vision 2050, we will also, whenever possible, recycle materials other than metal, such as concrete, glass, plastic, and paper. The waste plastic and paper that have deteriorated too much for recycling can be reused as fuel for producing electricity. Recycling these materials will consume less energy than production from natural resources, though the savings will be smaller than in the case of metal. Still, through recycling, we should be able to reduce the energy consumed in production of non-metal goods to 80% of the levels in 1995.

By estimating the relative quantities of metal and non-metal goods that will be produced in 2050, we project that through these increases in the rate of recycling of basic materials, we could reduce the energy used in production of goods to 70% of the energy used in 1995.

The second way to reduce energy consumed in the production of basic materials is to improve technologies for manufacturing both from natural resources and recycled materials. Improving the efficiency of today's most advanced technologies by 30% is a reasonable target, and achieving that would reduce the energy consumed in manufacturing to 70% of what it is today.

Differences in Energy Efficiencies Among Countries

The third way in which we can reduce the energy consumed in production of basic materials is by transferring technologies from countries having the most advanced production processes to countries using old energy-wasting technologies. We will see here that the effects of technology transfer are both large and reliable.

Until this point, the numbers and graphs in this book showing the efficiencies of "making things" and of generating electricity with fossil-fuel fired power plants have been mainly for technologies in Japan. While this is in part because it has been easier for me to get information on technologies from my home country of Japan, it is also the case that many of the technologies in Japan are the most energy-efficient in the world. Thus using the figures from Japan has given me a chance to introduce examples of the highest levels of energy efficiency. The amount of energy consumed in production varies greatly, depending on the country in which the

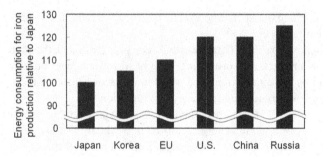

Fig. 7-3: Comparison of unit energy consumption rates of iron production in major iron producing countries relative to Japan (Courtesy of Japan Iron and Steel Federation)

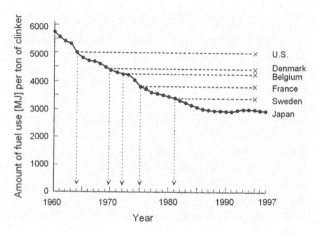

Fig. 7-4: Unit energy consumption rates for the Japanese cement industry from 1960 to 1997 and the positions of various countries in 1995 (Courtesy of Japan Cement Association)

goods are made. For example, in figure 7-3, we see that the energy consumed in making one ton of steel from iron ore varies as much as 25% – from Japan, with the highest efficiency, to countries with lower efficiency, such as China, Russia, and even the U.S.

Figure 7-4 shows a graph of how the amount of energy consumed in Japan to make one ton of cement changed from 1960 to 1995. The graph also shows comparisons with the energy efficiencies for cement making in other countries. Between 1960 and 1995, the energy consumed in making one ton of cement in Japan dropped by half. In comparison to Japan, most other countries used a much greater quantity of energy in 1995 to produce one ton of cement. The U.S., in particular, stands out – using 1.7 times more than Japan did at the time the graph was compiled. Thus the energy efficiency of 1995 U.S. technology in making cement corresponds to that of Japan in 1964.

This difference in energy consumption is a simple reflection of the rate at which each country has introduced new technologies to conserve energy. In the case of cement, the difference shows to what extent energy-saving technologies such as "suspension preheating" and more recently "new suspension preheating" have been introduced. These technologies thoroughly recover heat when coal is burned at high temperature, using the high-temperature gas emitted from the calcination furnace to preheat the powdered coal fed into the furnace. The term "suspension" comes from the way that the coal powder is suspended in the air by the high-temperature gas coming from below when the heat is recovered. In 1995, 87% of plants making cement in Japan used "new suspension" technology, and the remaining plants were all equipped with "suspension" technology. In the U.S., the number of plants using either technology was almost zero.

Just by introducing technologies of "suspension preheating" or "new suspension preheating" – already in use in Japan – to cement-making in the U.S. and the EU, we could conserve energy. And the investment capital for such a retooling could be recovered within a few years. The only reason these technologies have not been introduced already is the current unfavorable relationship between investment and return in many countries.

Not only is it possible to cut energy use through technology transfer, but doing so yields higher investment efficiency when considered at a global level. All that is necessary for benefiting from technology transfer is to come up with the capital needed to retrofit existing plants for the new technologies. However, improving cutting-edge technologies requires large investments in research and development. And because when we develop technologies for reducing the emissions we tackle first those emissions that are easiest to control, the return on investment in such research will inevitably decrease over time. Though there is still some potential for improving technology to increase energy efficiency in the production of basic materials and goods, the gaps between today's most advanced technologies in production and the theoretical limits are not as great as is the potential for improving the efficiency of transportation, homes, and offices.

In some countries today, the use of energy is particularly inefficient. In the countries of the former Soviet Union, for example, despite a much lower standard of living, the amount of energy consumed per capita is about the same as in Japan. Consequently, by improving technology, those countries should be able to achieve the same standard of living now enjoyed in developed countries without increasing current energy consumption at all.

By bringing energy efficiency world-wide up to the level of the most advanced current technologies, we could reduce energy consumption by as much as 30%, thereby reducing the energy required to "make things" to 70% of current levels.

If we combine the effects of the three ways for reducing energy consumption in production of materials – recycling, improving technology, and transferring technology – you can see that it would be possible to reduce the amount of energy consumed in producing material goods to $0.7 \times 0.7 \times 0.7$, or about one third.

In summary, in Vision 2050 we will cut energy consumed in transportation to one quarter of current levels, energy consumed in homes and offices to one quarter of current levels, energy consumed in "making things" to one third, and energy used in other industrial sectors – such as construction, home appliances, and heavy machinery – to one half. When the relative amounts of energy used in each of these sectors are taken into account, the resulting savings would mean that we would be using less than a third of the energy we are consuming today. In other words, by doing the things outlined above, we could – as stipulated by Vision 2050 – triple the efficiency of energy use.

The Potential for Energy Reduction

You may have noticed that the reduction goal for transportation and maintaining homes and offices is considerably larger than that for "making things." Furthermore, in reaching the reduction goals in "making things," the savings that we have projected will come through improvements in technology is just 30%, with the remaining savings to come from recycling and technology transfer.

The theoretical potentials for reducing energy in making steel and in driving automobiles are different. As demonstrated in Chapter 5, in making iron from iron ore, we must use energy to displace the oxygen atoms bound to the iron in iron ore. Currently, this energy is equivalent to one third of the total energy used by an iron mill. We can consider this energy to be internal energy "embodied" in the pig iron produced, or to put it differently, the pig iron produced by the iron mill inherently contains energy equivalent to 200 kg of the 600 kg of fossil fuels that are currently used to produce one ton of iron. Only the remaining two thirds of the energy is "lost" in the process, and so the reduction potential in the making of iron is just 400 kg of fossil fuel per ton of iron.

We saw in Chapter 3 that the theoretical minimum energy needed for transportation is zero. This means that the reduction potential for driving automobiles is the entire amount of fuel used. Therefore, it is clear that the reduction potential for transportation is much greater than the reduction potential in the production of iron.

In addition, energy constitutes a smaller fraction of the total cost of "making things" than it does for transportation or running homes and offices. Here's why: until now there has not been a strong demand for energy efficiency in products such as refrigerators, air conditioners, and cars. Instead, design and performance have been more important in giving a competitive edge to such products. The cost of electricity for a typical household, on the order of $1,000 per year, has not been a strong stimulus for energy conservation.

On the other hand, consumer preferences are not an issue in the design of processes for "making things." The consumer is usually not interested in or concerned about the process used to produce the iron used in a car as long as the performance of the car is not affected. Therefore, controlling energy costs (along with improving efficiency in converting raw materials into products) has long been a large factor in reducing the cost of manufacturing products. For this reason, manufacturing companies have invested heavily in R&D and facility improvements, striving to increase energy efficiency in order to maintain their competitive edge.

In summary, Vision 2050 places a higher expectation on energy conservation in "daily life" activities such as transportation and running homes for the two reasons: 1) the gap between the present energy use and the theoretical limit is larger in the case of "daily life" activities, affording more opportunities for conservation, and 2) most efforts at energy conservation until now have been in the arena of "making things," which means that the yield on efforts for further energy conservation are likely to be minimal.

(2) Construction of a Material-Recycling System

Metal and Concrete

In 2050, we will probably still not have made a complete conversion from fossil fuels to renewable energy, and human artifacts will probably not have reached a state of complete saturation. However, by 2050 we need to create a launching platform that aims us in the direction of an ultimate state of complete conversion from fossil fuels and saturation of human artifacts by the end of the next century.

Let us take a look at the lifecycle of iron in Vision 2050. As the accumulated iron nears saturation, the amount of iron ore that is reduced will decrease, so the total amount of iron accumulation of 35 billion tons, which might have occurred if the present rate of production of 900 million tons per year from iron ore were continued unabated, will not be reached. The amount of iron accumulated by 2050 is predicted to be about 30 billion tons. If the average product life is the same 30 years that it is today, then one billion tons of scrap will be generated each year. We will use this scrap, minus a small amount of waste that is thrown away in garbage dumps, together with 200 million tons of iron ore as raw material for new iron. Thus, the world will produce 1.2 billion tons of iron per year in 2050, but 85% will come from scrap. Let us consider that in 2050 the global average consumption of coal per ton of iron will be 500 kg in the case of iron made in blast furnaces due to advances in technology that reduce coal consumption by 100 kg, and 150 kg in the case of production from scrap. The total coal consumption for iron production will then be about 250 million tons per year. Even though the amount of production will have not changed, the amount of coal consumption will become almost one third the present amount, which as we saw in figure 1-6 is about 700 million tons per year. This is a concrete example of the effect of the three-fold increase in energy efficiency for iron production due to recycling, technology transfer, and technology development that we discussed in the previous section.

After the quantity of iron accumulated in the cities, roads, and other durable products has reached about 39 billion tons, there will be enough scrap generated each year so that all of the iron that is needed can be produced from scrap. Furthermore, when all fossil fuel use is completely replaced by renewable energy sources, all of the energy for producing the iron from scrap will be supplied by renewable resources. This is the ultimate form of the lifecycle of iron that we should aim to realize in the 22nd century. Vision 2050 is different from this ultimate form, but compared to the present, it is much closer. The differences are that there is still some need for extraction of iron ore and fossil fuels, and also a part of the waste iron still winds up in garbage dumps. In fact, there will always be a fraction of waste generated that is useless scrap, unfit even for recycling. The next chapter will take a look at how we will treat this small amount of waste material that is not recycled in even Vision 2050.

Next, let us consider the lifecycle of concrete. Although waste concrete, produced for instance from the demolition of buildings, is currently used in low-grade applications such as road paving materials, we have see that as the amount of waste material grows, the fraction that is thrown away in garbage dumps will increase. In 2050, it is predicted that, like iron, the accumulation of concrete will reach three to five times the current amount, and the amount of waste concrete will grow in proportion. In fact, from 1995 to 2007 the worldwide production of concrete has nearly doubled, mainly due to increased output in China. To prevent the earth from being buried in waste concrete, it is necessary to construct a nearly perfect recycling system for concrete. One way would be to develop a technology for the regenerative pulverization of concrete, where waste concrete is pulverized into a sufficiently fine power so that the raw material for making cement can be recovered.

Paper and Plastic

Compared to iron and concrete, materials such as paper and plastic, which are used in artifacts with much shorter product lives, will saturate at smaller accumulation amounts. Therefore, for these materials it should be possible to arrive at a condition close to the ultimate recycling society even by 2050.

Today, already about half of the paper that is used is recycled, and most of the remainder is thrown away in garbage dumps where it is eventually released to the atmosphere as CO_2. In 2050, by increasing the recycle ratio, two thirds of used paper will be fed as raw material into the process of making new paper, and the remaining one third will be used as fuel. We will need to harvest a sufficient amount of trees to replace the one third of the waste paper that is used as fuel in order to maintain the annual production rate of paper, and we will replant trees at the same rate that they are harvested. We will develop paper manufacturing technologies by 2050 that make it possible to produce a ton of paper with just 200 kg of carbon – a 70% improvement over the present technology level. One third of the used paper will be used as fuel in papermaking, and by converting to carbon units, we find that this is exactly enough energy to produce new paper from the other two thirds of the used paper. Looking at this lifecycle of paper as a whole, we see that forests are being replanted and there is no consumption of fossil fuels, so the CO_2 concentration will not be increased. This is an example of a perfect recycling lifecycle.

The future state of technologies for manufacturing chemical products, as represented by plastic, is difficult to predict. Although currently almost all plastics are produced from oil, as long as there is a source of carbon and hydrogen, it is possible to synthesize plastic from raw materials other than the oil. One possible alternative to oil as the raw material for making plastics is biomass. For example, the process of making various chemical products from carbon monoxide and hydrogen synthesized with biomass as the raw material, called C1 chemistry from the fact that carbon monoxide is a feedstock with one atom of carbon, is technologically feasible even today. Also, researchers are developing ways for growing plants that produce the raw materials for plastics through biotechnology.

In all likelihood, society will continue to require a broad range of high performance chemical products. We must construct a system to supply society with materials that can meet these requirements, that can stand up to recycling, that have excellent combustion efficiency when they reach the end of their life cycle and are used as fuel, and that present no threat of releasing toxic substances such as dioxins or endocrine disruptors throughout their entire lifecycle.

In summary, each of the major basic materials – metals, ceramic materials, paper, and plastic – show Vision 2050 lifecycles with their own special characteristics. However, in comparison to the present, each of the lifecycles we have seen here contributes to the reduction of the factors that are interfering with the circulations in the biosphere – the amount of CO_2 emissions, the amount of waste material disposed in landfills, and the amount of underground resources that are extracted – and therefore each one can form a part of a sound intermediate stage towards the ultimate goal of a perfect recycling society.

(3) Development of Renewable Energy

Aim to Double the Present Amount

As shown in figure 7-1, in Vision 2050, we will reduce the use of fossil fuels to three quarters of what it is today. This reduction is absolutely necessary in order to control global warming from CO_2. In order to achieve this reduction while still providing the same amount of energy as today, we will introduce a supply of energy equivalent to one fourth of the current consumption of fossil fuels through the development of renewable energy. Because the renewable energy resources that we introduce will not emit any CO_2, CO_2 emissions will be reduced by the amount of renewable energy that is introduced: that is by 25%.

Hydropower already supplies 5% of the global energy demand. The conversion efficiency to electricity for hydropower is high, so as long as we take care not to cause other environmental problems such as the submersion of large regions of land, as an energy resource it is ideal. Consequently, in Vision 2050 we will develop new hydropower at a scale similar to the present. We will develop applications with electric power demand for that hydropower such as aluminum production close by the hydropower plants, and we will also locate hydropower plants so as best to meet to the increase in electric power demand in developing countries. The development of hydropower in Iceland as a source of power for aluminum production could be a good model for this process.

Another important issue that is related to material circulation is the problem of what to do with the biomass that is currently being thrown away. Used paper is one example that we have already looked at, but other kinds of biomass are also thrown away in large amounts. In fields where the autumn harvest has been finished, we sometimes come across the picturesque view of straw being burned in the fields; however, this is essentially just the same as "burning oil fields." It has been estimated through conversion to carbon units that about two billion tons of unused residual biomass is generated from agriculture and forestry worldwide. Even if we are only able to utilize half of this biomass effectively, we could still substitute for the equivalent of one billion tons of fossil fuels.

If we construct an efficient and effective collection and reuse system for municipal waste, which is something that we need now anyway, or for residual materials from agriculture and forestry, which we have seen could be a large resource, such a system would be usable almost immediately. Also, we could create biomass energy plantations using available land such as fields that are lying fallow, to develop another 900 million tons of biomass production, or 15% of the fossil fuel consumption in 1995. Of course we must be careful not to reduce the world production capacity of food grains, and having a shared vision such as Vision 2050 should help us to do that, by making the tradeoffs involved in each choice clear to all people concerned.

It should be possible to develop enough solar power to produce electricity equivalent to 200 million tons of fossil fuels, or 3% of the 1995 fossil fuel consumption. We could also pursue the development of wind power and geothermal power, taking care not to cause other environmental problems. In Vision 2050 we will act to advance the development of all kinds of renewable energy by mapping out the improvement of energy technologies through scientific research and by building up a manufacturing infrastructure for enabling these technologies to spread throughout society.

Summarizing the above, we will aim to achieve the new development of hydropower equivalent to 5% of the current fossil fuel consumption, biomass such as agricultural and forestry residuals and municipal waste equivalent to 15%, solar cells equivalent to 3%, and the equivalent of about 2% of current fossil fuel consumption from other renewable energy sources such as wind and geothermal. This gives us a total of 25% of 1995 fossil fuel consumption, or 1.5 billion tons of fossil fuels, that will be substituted by renewable energy sources in Vision 2050.

The fraction of energy generation made up by solar cells in Vision 2050 is just 3%, which is considerably less than that of biomass and even of hydropower. Why can we not aim to achieve more? The reason is that, even if the technology is achieved, we will probably not be able to develop the total amount of energy supplied by solar cells in 2050 to a scale that greatly exceeds 3% of the total energy demand. As a general rule, it takes time to go from the development of an energy technology to the actual widespread penetration of that technology into the market. In particular, solar cell technology has the characteristic of a large initial investment cost and almost zero running cost. After the cells are manufactured and put into place, there is essentially no additional cost, and eventually the cells will pay for themselves. However, the initial cost to make and install the cells is still formidable. On the other hand, although investment costs for biomass energy systems are low, costs are incurred when collecting the biomass and transforming it into an easy to use form of energy such as electricity. Furthermore, while it is expected that commercial solar cells may reach conversion efficiencies of as much as 40%, we saw in Chapter 6 that the limit for biomass is about 5%. Consequently, while biomass is a technology that can be used right now due to the low investment cost, it has considerably less potential for being a major player in the future than solar cells. This is one important way in which the characteristics of different renewable energy technologies are different.

We might begin to create a solar power infrastructure by installing solar cells on the roof tops of city buildings and then expand the development of solar power into other applications. Through the cycle whereby increase in demand drives progress of technology, technology will improve, and gradually a solar cell infrastructure and industry will become established that will prepare the way for a much larger contribution of solar power in the second half of the 21st century. More generally, in Vision 2050 we need to plan out what kind of human artifacts we should begin to accumulate in the social infrastructure. Because solar cell technology is characterized by high initial costs followed by near zero running costs, in exchange for not expecting an excessively large contribution in Vision 2050, we must work to set the stage for a greater contribution to come later.

Towards a Perfectly Recycling Society

In the previous sections, we have seen how it is possible to move towards the establishment of a completely sustainable, perfectly recycling society from the second half of the 21st century using Vision 2050 as a road map. Moreover, rather than just being sustainable, it will be a society that lets us expand our lifestyles even further. The global amount of energy consumption will be almost the same as it is today at the point when this intermediate target of Vision 2050 is reached, and that is just about one ten thousandth of the total amount of the energy that shines down on the earth from the sun. Both biomass and solar power have the potential to provide more than enough energy to meet our energy needs today, so there is plenty of room to increase our energy use through the development of these resources. What we need to do in Vision 2050 is to move towards a breakaway from fossil fuels and spur on the acceleration of the introduction of renewable energy and recycling technologies.

Through well-planned development of technologies for a sustainable earth, we will eventually be able to supply much more energy for human consumption than we do today. For example, by exploiting just two ten thousandths of the sun's energy, we would be able to use twice as much energy as we do now. Electric vehicles that run completely clean, houses that are equipped with comfortable heating and cooling systems, beautiful and healthy oceans and forests that are located right next to large cities, all maintained using renewable energy – this vision of the future is not just a dream.

Chapter 8
How Will Technology and Society Work Together?

The previous chapter has presented Vision 2050 as a road map to a sustainable earth. In earlier chapters, we have tried to demonstrate that this vision can become a reality only if scientists, industry leaders, and policy-makers around the globe work together to develop, implement, and share technologies for sustainability. However, this vision also requires the support and participation of the general public. For example, it would be impossible to make new products from waste materials without the cooperation of local citizens in recycling and without the creation of an infrastructure for separating and collecting garbage. In this chapter, let's consider the problems that arise at the point of contact between technology and society and how we can address these problems.

1 Forming a Total Infrastructure for Circulating Materials in Society

The Importance of Separation in Garbage Collection

Depending on how it is collected, household waste can be either a resource or a burden on the environment. We have seen that waste paper can be used as the raw material for paper, and waste plastic can be raw material for plastic. But although a mixture of paper and plastic might be useful as a source of energy, the mixture cannot be used as a raw material. Furthermore, if food waste is mixed in with the paper and plastic, the mixture cannot be used even as an energy source. If, for example, we tried to burn a mixture of paper, plastic and food waste to generate electricity, heat would be lost in vaporizing the water in the food waste. As a result, the efficiency in generating electricity would barely exceed ten percent, and that much electricity would be used up just operating the plant.

Hiroshi Komiyama and Steven Kraines
Vision 2050: Roadmap for a Sustainable Earth.
© Springer 2008

Many of the recycling systems in operation today are not designed to produce high quality materials. Producing high quality materials from recycling requires an integrated system that includes separation during garbage collection and possibly even the redesign of products to make it easier to separate component materials during recycling. If we mix an artist's various paints together, all the bright colors turn to grey. Likewise, if we do not separate the different colors of glass we use in daily life when we recycle, the color of the recycled products will approach a dingy shade of grey. If we want to maintain a variety of different colors of glass in recycled products, we must develop an adequate collection system and recycling technology. However, the choice of whether to take on the trouble of a complex separation system or to make do with a single color of glass is a choice that the citizens of each society must make.

When I was in Switzerland, I noticed large metal containers set out in various locations for recycling. The Swiss separate their glass, paper and plastic, and put them in those containers. Glass is separated into three different containers by colors: clear, green, and brown. After collection, this glass is pulverized and impurities such as metals are removed. The pulverized glass is then melted down and reshaped into new products. There is no need to divide the glass by size or shape, but to make recycled products of a particular color, each color of glass must be collected separately.

Another way to recycle glass products is to reuse the glass product as it is. This is the case with glass beer bottles in Japan, a case often cited as an exemplar of recycling. It takes additional effort, but this form of recycling consumes even less energy while maintaining the color and quality of the original product.

There are many ways to recycle. And to maintain our present lifestyle, each society must set up a recycling system that combines different recycling methods in the way that best meets its needs. For example, a recycling system for glass might reuse beer bottles and other standardized glass products as they are, separate the remaining glass into several different colors that are melted down and formed into new glass products, and use natural resources to manufacture only the top quality products such as flower vases and ornaments made from lead crystal.

A Minimum Amount of Waste Emissions

The scenario for glass described above – where if waste materials are not separated, all recycled glass products will be a dingy grey – applies to other materials, such as metals and plastics. We have seen how, as a result of the "saturation of human artifacts," there will eventually be enough scrap to make all of the metal required by society. Consequently, a recycling society will recycle metals over and over. But if nothing is done about the impurities and additives in the metals, they will accumulate with each round of recycling until only low-quality recycled metals will remain.

In plastic products today, aesthetics, strength, and sealing properties are obtained by mixing different types of polymers or laminating different kinds of plastic in layers. However, if all plastics are recycled together without being separated into different types of polymers, we cannot expect to produce the same high quality in recycled products. At best, we will be able to use recycled plastic only for things like planters, park benches, and the filling material for car seats.

In a recycling society, to prevent the quality of materials from degrading, we must, in addition to separating waste material during the collection stage, do all that we can to prevent mixtures of different materials from forming in the first place. As you will discover in the next section, we can prevent mixtures by standardizing products and by developing new materials that perform at a high level without being mixed with other materials. Still, no matter how much we work to design materials and products to avoid mixtures and no matter how much we invest in a good separation system for recycling, some amount of impurities is bound to get mixed in. Therefore, we also need to develop technologies to increase the purity of recycled materials to the level of materials currently produced from natural resources by removing impurities with just a small amount of energy.

Finally, although we should be able to collect most waste materials at a level of purity sufficient for recycling, there will inevitably be some waste that cannot be recycled, such as heavily rusted metal or rotted out concrete. And we will probably continue to obtain some materials from natural resources, particularly for products requiring the highest purity, such as lead crystal. But as long as the minimum amount of waste material that is too degraded to be recycled and the amount of natural resources needed for top quality products do not exceed the long-term regenerative processes of the earth, a society that has reached a saturation of human artifacts can still be made sustainable while maintaining the quality of the materials used in society as a whole.

Product Design and Standardization

To make a recycling society workable, we will probably need to regulate the design of many products. For example, a large percentage of drink bottles made of polyethylene terephthalate plastic, otherwise known as PET, are currently recycled. But because the caps of these bottles are often made from a different plastic or even from metal, a high level of impurities remains in the recycled plastic. We could require that the materials used for the caps of PET bottles be limited to PET. Similarly, while iron and aluminum can be recycled efficiently if collected separately, if they are mixed together, it is much more difficult to recycle them. Therefore, it might be appropriate to prohibit mixtures of iron and aluminum in a single product.

Developing new material technologies could make materials easier to separate for recycling. For example, we could invest in the development of single polymers having nearly the same high performance features as present-day plastics, which are made of a mixture of different polymers, a mixture that is difficult to recycle. Another promising example is developing new substances to treat the surface of metals such as iron and aluminum, substances that vaporize when the metal is melted. For example, the zinc used for the surface treatment of iron vaporizes when the iron is melted down for reuse; therefore, the accumulation of zinc as an impurity is extremely small. The zinc can be easily separated from the iron once it vaporizes, so the zinc can also be recycled. On the other hand, tin, which is used for the same surface treatment, does not vaporize at the melting temperature of iron. Therefore, it is necessary to find another way to remove tin impurities from recycled iron. Although it may be difficult to develop these technologies, it is certainly possible.

Standardizing the specifications for products and materials would also make recycling easier. In the automobile industry, specifications for the additives in steel for body parts or the composition of windshield glass differ from manufacturer to manufacturer. Although it is possible for the current production processes to create materials from natural resources meeting all of these different specifications, to recycle material from the scrap that is produced would take an excessive amount of energy. But by standardizing these specifications, we could make recycling much more efficient.

Choosing the Optimal Scale

One fundamental principle upon which our infrastructure for material recycling must be based is the "scale effect" of industrial manufacturing. As we showed in Chapter 5, in general recycling consumes less energy than producing goods from natural resources. But if we were to collect glass, pulverize it, melt it down, and form it into new products in every city district or town, the small-scale of these operations would result in an inefficient use of energy. There are many situations like this where, if the scale is small, the efficiency will be low.

If glass is melted down in a small furnace, a large quantity of excess fuel will be consumed as heat is lost through the furnace walls. In a large furnace, heat escapes less easily, so we need only enough fuel to supply the heat for melting. The critical factor here is the surface area of the furnace divided by the volume, called the "specific surface area." The volume of a regularly shaped container such as a sphere increases at a faster rate than its surface area. Therefore, the specific surface area is smaller for a large furnace than a small one. A small specific surface area means less heat loss through the furnace walls. Also, the cost of equipment like furnaces and reactors per unit production capacity is generally proportional to the specific surface area. This is so because, while the amount of material used to build a furnace is proportional to its surface area, the capacity of the furnace is proportional to its volume. Therefore, a large furnace, with its greater capacity per unit of construction material, is not only more efficient to operate but also more economical to construct.

Process industries – such as glass factories, iron and steel mills, and petrochemical plants – have continued to increase the size of their plants to capitalize on these scale effects for energy efficiency and equipment cost. The same kind of scale effects apply to the production of materials through recycling. As a rule of thumb, the size of present-day plants for manufacturing a particular material is probably a reasonable target for a plant that recycles the same material. For example, irrespective of whether glass is created from natural resources or from recycled materials, the energy consumed during the melting and shaping processes will decrease if the scale is increased.

However, there are some situations where a larger scale may not be better. How to handle food waste is one major problem we must address to achieve the comprehensive circulation of materials required for a sustainable society. Food waste can impede recycling by being a source of contamination in the material to be recycled, by causing formation of toxic chlorine-based chemicals from the combustion of the chlorine in salt, and by reducing the efficiency of generating electricity due to the high water content. Food waste has a high water content, so we could collect and process this waste more efficiently if we could remove the water. It is easier and more efficient to remove water from food waste on a small scale because when the waste is divided into small amounts, it has a larger specific surface area. At the household level, water could be easily removed from food waste by drying it in a solar-heated compost box, spin drying it in a disposer, or using some other small-scale method. And if the water is removed where the food waste is generated, we will save energy in transporting the dry food waste to be recycled because it is lighter and easier to handle.

Heat pumps are another example where sometimes better efficiency can be attained on a small scale. The efficiency of small scale heat pumps, such as air conditioners for home use, is not necessarily less than the efficiency of those used in large buildings. We have seen that one of the main factors determining the efficiency of a heat pump is the efficiency of heat transfer between the heating and cooling units and the air. Using lots of small indoor and outdoor heat pumps, such as home air conditioners, results in a larger area for transferring heat. Therefore, it could be at least as efficient to use individual air conditioning units for each room in your home as to use a central unit, particularly when you consider that a central unit must distribute the heating and cooling throughout your house, resulting in loss in the ventilation system as well as needless heating and cooling of unused rooms.

The point to keep in mind is that processes requiring area, such as drying and cooling, can be carried out on a small scale, but processes requiring volume, such as melting and chemical reactions, should be done on a large scale. In other words, when we want to minimize the loss of heat from a process, we should do that process on a large scale, but when we want to maximize heat transfer, it can be advantageous to do that process on a small scale. We must adopt this as a fundamental principle when we formulate a comprehensive plan for material circulation.

A Network System for Biomass Collection

Constructing the infrastructures in society to facilitate material circulation is important in other areas besides recycling. For example, to effectively use the residual by-products from agriculture and forestry as biomass to produce energy, we need a collection system. And to combust this biomass efficiently for generating electricity, we need a drying system.

Because drying requires surface area, it is inefficient to dry the biomass residuals on a large scale, after a huge quantity of residuals has been collected at a single location. It is better to use the energy of the sun to dry the residuals where they are produced – at the farm or lumber mill. Then we can collect the residuals in stages, starting with an initial drop-off point to which the producers of biomass bring the residuals over a distance similar to the distance they now transport harvested goods. From these initial drop-off sites, the dried biomass can be collected and carried to middle-level collection points, and so on. Transporting loose straw and husks wastes energy because of the bulkiness of the material. Energy could be saved by compressing the residuals into solid blocks that take up less room and are easier to handle. The optimal place to install equipment for compressing is probably the middle-level collection points. Finally, we must make the power generation plant at the final stage large enough because generating electricity by burning biomass is a process that benefits from a large-scale operation.

Dried biomass compressed into solid blocks, called "RDF" for refuse derived fuel, has a fuel value comparable to coal. Moreover, the content of pollutants, such as sulfur, is typically lower in biomass fuels than in fossil fuels. Judging from current levels of technology, if we could collect biomass on a sufficient scale, it should not be hard to convert it into convenient forms of energy, such as electricity or vehicle fuel.

But it is vital that such a system be constructed with the assent and understanding of the farmers and other participants, regarding factors such as the modes of transportation, the construction and layout of collection points, and the distribution of costs. The borderline between effectively harnessing a huge amount of natural energy and creating just another "burning oil field" lies in collaboration.

Production in the 20th century was a one-way flow from natural resources to human artifacts supplied to the market. Because of that one-way flow, technologies were developed independently for each plant. However, if we are to make the transition to producing goods from recycled artifacts, technology must be shared throughout a large social system that includes the standardization of human artifacts, the design of systems for collecting waste materials, and the development of methods for recycling. Because a society that efficiently recirculates materials depends on collaboration, a good relationship between society and technology is essential.

2 Making the Market Work for Sustainability

Can We Leave Things to the Invisible Hand of the Free Market?

After the end of the Cold War between the capitalist world and the communist world, the debunking of planned economies following the collapse of the Soviet Union created the impression that market principles, or "the invisible hand," had prevailed over all other economic systems. In Japan, people have been clamoring for deregulation for years. It often seems as if all our problems would be solved if we just eliminated all regulations.

However, in a situation where the world's population as a whole must respond with long-term vision to the environmental and energy problems threatening to undermine the foundations of civilization, can we leave the decisions solely to the "invisible hand" of the market? Probably not. As long as corporations act on short-term outlooks, the principles of the free market will never attain the level of cooperation required to meet the large-scale, long-term problems of sustainability. One problem is that many of the negative consequences of human activities, such as CO_2 emissions from transportation, are not properly priced for the market mechanisms to work. Recently, much concern has been raised about the environmental costs of purchasing goods produced in countries far away. There are many similar examples where excessive burdens on the earth occur as a result of mismatches between prices and environmental costs.

If we look at the global circulation of iron, the problem becomes clear. In Japan, at the start of the economic boom in the 20th century, iron scrap was imported. However, as a result of rapid economic growth, human artifacts made of iron accumulated, the amount of scrap generated domestically increased, and in 1992, export of scrap iron surpassed import. Currently, Japan exports 7.6 million tons of iron scrap, but it still imports 180 thousand tons. In the U.S., the situation is even more extreme. Since the 1950s, the U.S. has been a net exporter of iron scrap, but since the 1970s, the U.S. has also imported a substantial quantity of iron scrap. In 2007, although the U.S. exported 14.9 million tons of iron scrap, it also imported 4.8 million tons.

So why is it necessary to both export and import iron scrap instead of just exporting the difference? The reason is related to the nature of iron products in the U.S. and Japan. In the U.S. and Japan, demand for high-performance products is large, so high quality iron scrap such as unused cutoffs is needed. On the other hand, iron scrap generated from human artifacts that have reached the end of their product lives is often rusted, may have bits of concrete attached to it, and contains a lot of different impurities, so it is not easy to use in high quality products where composition and minute structure must be precisely controlled. As a result, low quality scrap has become overabundant in the U.S. and Japan, so it is exported. In developing countries, there is still a need for structural materials that can be made from cheap scrap, so there is a demand for even low quality scrap.

But as you have seen in this book, eventually low quality scrap from human artifacts will be generated in much larger amounts in countries around the world, and a surplus of low grade scrap will occur worldwide. On the flip side, the demand for high quality scrap will increase as developing countries begin manufacturing more high-performance products, resulting in a shortage of high quality scrap. At that point, how will the iron and steel companies respond? If we stood in the shoes of the executives of those companies, we would inevitably choose to continue the reduction of iron ore in blast furnaces. Rather than tackling the troublesome task of processing low grade scrap to produce high performance products, it is more economical in the short run to use the high purity pig iron made from iron ore, which is still in plentiful reserve. It is clear from this example that if we entrust the production of iron to the invisible hand of the market without any form of regulation, the circulation of iron will not happen. To achieve a material-recycling society, the market must be influenced in such a way that recycling becomes economically advantageous.

Guiding the Market

In the previous chapter, we saw how manufacturing industries have achieved tremendous reductions in energy use during the last few decades. As a result, the fraction of the total cost made up by energy cost in Japanese industries is just 20% for the highest consumer of energy: the cement industry. For chemicals, iron and steel, and paper and pulp, the fractions are 15%, 14% and 6%, respectively. Therefore, the economic drive to invest in energy conservation is considerably reduced. As long as fossil fuels continue to be as cheap as they are today, it is probably not advantageous for industries to invest further in energy conservation.

On the other hand, the general public cannot be expected to develop energy-conserving habits on a large scale either with the price system as it is now. With a car that gets 10 km per liter, a motorist who drives 10,000 km per year and pays one dollar per liter (or $4 per gallon) for gasoline will spend a thousand dollars a year on fuel. If that person were to buy a hybrid car with 50% better fuel efficiency, the annual savings would be five hundred dollars. Hybrid cars today cost over five thousand dollars more than conventional cars with equivalent performance features, so it would take more than ten years of fuel savings to pay back the difference. Because most people own their cars for no more than ten years, there is little economic incentive to purchase energy efficient automobiles. As a result, energy efficient automobiles, such as hybrid cars, are purchased primarily by consumers concerned about the environment and not by consumers responding to market forces.

One way to influence the market towards energy conservation is to raise the taxes on energy. As shown in figure 8-1, more than 50 cents per liter of the cost of gasoline sold in Japan today is tax. Many other countries impose similar or even larger levels of tax on gasoline. On the other hand, the tax on gasoline in the U.S.

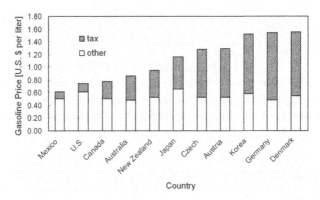

Fig. 8-1: The fraction of gasoline prices made up by tax in various countries (Data from International Energy Agency, Energy Prices and Taxes, 1st Quarter 2007)

is only about 13 cents per liter. This low tax rate accounts for most of the large difference between American and Japanese/European gasoline prices shown in figure 8-1.

Unfortunately, even the high gasoline tax imposed in Japan today is not enough to motivate people to purchase energy efficient cars for economic reasons alone. Therefore, more direct ways to tilt the market towards energy conservation are being considered. One approach, called the "top runner" method, places a tax on cars based on the amount of energy they consume, using the car with the lowest energy consumption rate as a benchmark. As a policy to promote energy conservation, this method makes sense. However, Japan's proposal to adopt a top-runner tax initially met with strong resistance from the EU, whose citizens tend to prefer cars with lower fuel efficiency. Only after several years of negotiations did the EU finally adopt a top-runner tax system.

Even in the EU, which took a leadership role in negotiating the control of CO_2 emissions at the Kyoto COP3 meeting, when discussions reach the point where policies directly affect domestic industries, national governments often are forced to change their stance. National self interest is often an obstacle to addressing global environmental problems: long term benefits to humanity can and do conflict with the short term interests of individual nations. But increasing the energy efficiency of automobiles is essential for meeting the Kyoto goals as well as for achieving a sustainable earth. So workable agreements must be adopted and enforced.

Similarly, the use of renewable energy sources will expand slowly if left to the forces of the free market. Many options for renewable energy require a steep initial investment. For example, installing solar cells to meet the electricity requirements of a single home costs about 20,000 U.S. dollars. Given current electricity prices, it would take many years for a home-owner to recover the investment costs through savings on electricity. However, we have seen that the energy needed to manufacture and install the solar cells can be recovered in two years, so from the overall perspective of conserving energy, installing solar cells is good. Different methods for adjusting the market to favor the introduction of renewable energy are being studied and applied. One example is the *Aachen* method in Germany, where electricity prices are raised by 1% and the added revenue is used to subsidize the development of renewable energy.

Another example is the Feed-in Law introduced in Germany in 1990. The Feed-in Law required utilities to connect small operators generating electricity from renewable energy technologies to the grid and to buy the electricity that they produce at close to the market price for final customers. This law was implemented to level the playing field of the energy market. The Feed-in Law was replaced in 2000 by the Renewable Energy Source Act. Under this act, the feed-in prices for electricity generated with renewable technologies are no longer linked to electricity retail prices; instead they are fixed for 20-year terms. Thus the generator is freed from the risk of being stuck with electricity it cannot sell. A sophisticated redistribution system ensures that the financial burden is evenly distributed to the end customer. The generator of renewable electricity is granted preferential access to the grid and has the right to be connected immediately. The feed-in prices offered to new installations will be lowered each year to take into account the decrease in investment costs for renewable energy as the technologies mature.

The "carbon-tax" method, whereby a tax is imposed that is directly proportional to the amount of CO_2 emissions, is another example of a mechanism used to guide the market towards sustainability. Already carbon-taxes have been put into effect in Sweden, Finland, the Netherlands, Denmark, Norway, Italy, and the United Kingdom. Carbon-taxes direct the market towards energy resources that have less carbon, many of which are renewable energy technologies. Results of a computer simulation reported by the National Institute for Environmental Studies in Japan show that by introducing a carbon-tax of 30 dollars per ton, Japan would be able to meet the Kyoto agreement.

Yet another example of how to moderate the force of free markets is the EU Emissions Trading System, initiated in 2005. This system is the world's largest tradable permits program, applying to approximately 11,500 installations across the EU's 25 member states. Many studies are being conducted on different aspects of this trading system, including efficiency and equity in distributing permits, implications of economy-wide programs versus regional ones, mechanisms for handling price uncertainties, different forms of targets, and issues in compliance and enforcement.

The development of ways to reform economic and political systems is outside the scope of this book. However, if adequate policies and guidelines are adopted, Vision 2050 is definitely within our reach.

3 Projects for Vision 2050

To successfully introduce recycling systems and renewable energy, we must develop large-scale social infrastructures. Those infrastructures must be based on application-oriented research and draw on a wide range of ideas for creating a sustainable earth. Because these society-encompassing infrastructure systems must transcend the frameworks of industry, the development of such systems cannot be left to individual companies. Instead, we must turn to other institutions in society, such as governments, international agencies, non-profit organizations, and universities, to lead these development projects. These institutions must collaborate with companies in planning, promoting, and implementing the society-encompassing infrastructure projects needed to create a sustainable society.

To achieve Vision 2050, what kinds of projects do we need?

Design of Giant, Complex Systems

As one example of a society-encompassing system for Vision 2050, here is a hypothetical design for how we might establish a material-recycling society.

First, for each basic material, we must design a system for circulating the material that limits the degradation of quality during circulation as much as possible. For iron, we might design an overall framework that includes the separation and collection of iron scrap generated when products such as buildings and cars reach the end of their product lives, a recycling process that removes as many impurities as possible using a reasonable amount of energy, and an information system for monitoring and communicating the quantity of recycled iron that can be produced at each level of quality. But when we try to implement this system, we will discover that we will not be successful if we limit this design to only the iron and steel companies. Problems will arise, such as how to coordinate with other industries including construction and automobile manufacturing, how to induce people to separate the garbage they throw out, and how to arrange the collection and transportation of waste materials. Even after we resolve these problems, we must trace how the primary material, iron, will circulate in society, and estimate how the additives and impurities such as phosphorus, copper, zinc, tin and nickel will be distributed in the various iron products.

Next, we must design a similar process for aluminum, cement, plastics, and all the other basic materials. Manufactured products are usually composed of many materials, so adjusting the amounts of different materials used in each product will be necessary. For example, we must regulate the use of substances that impair the recycling of high quality iron. This regulation must include even additives in other materials used in the product together with iron. For example, if glass is used as a surface coating for a steel car fender and that glass contains copper, then when the fender is melted for recycling, the copper will mix with the iron. Also, for heavy, low cost materials such as concrete, reducing transportation costs is essential, so we must plan where and how to separate concrete from other materials to minimize cost and maximize efficiency. We must design specifications for products, methods for recycling, and methods for collection, and these methods must be coordinated in such a way that few conflicts arise. Furthermore, we must map out a scenario showing how we will convert those specifications and methods into a functioning reality. In particular, we must decide when to use regulatory mechanisms, when to use subsidies, and when to rely on the free market.

This kind of material-recycling society is a much more complex system than today's society of mass-production / mass-generation-of-waste. We cannot hope to create such a complex system just by thinking up catchy slogans. We will need vision and strong leadership together with opportunities where the various constituents of society can orchestrate their collective efforts to make a sustainable society. We must bridge the communication gap between different stakeholders in society and create design tools for helping those stakeholders to fine-tune the overall system by communicating their ideas and their needs. We will look at these challenges in the last section of this book.

The design of a material-recycling society is one project we must undertake right away to reach the goal of Vision 2050. On the other hand, even though we do not expect technologies such as solar cells to make large contributions by 2050, we must encourage their research and development now. Technologies not expected to be widespread until after 2050 do not have immediate economic payoffs, so they cannot be simply entrusted to the free market. Instead, they must be nurtured through the collective will of society.

A Large-Scale, High-Efficiency Manufacturing System for Solar Cells

In 1998, the number of solar cell arrays that had been installed on roof tops in Japan was about 10,000. By 2007, the number had increased to more than 400,000. On average, each array for home use has a capacity of about 3.5 kW, so the total peak power generation capacity is 1,400,000 kW. However, this is the amount of power generated when sunlight is strongest. To compare the power generation capacity of solar cells to that of thermal power plants, we need to account for both the daily variations and the seasonal variations of sunlight. The average power generation of solar cells calculated in this way decreases to about one tenth of the peak generation capacity. Therefore, considering that the total electricity generation capacity in Japan today is about 200 million kilowatts, less than a thousandth is provided by solar cells.

By installing solar cells on all of the roofs in Japan, it would be possible to meet over 20% of the current demand for electricity, or 6% of the total energy demand. However, even if the annual production capacity of solar cells could be increased to one hundred times the current capacity, it would still take more than one hundred years to produce that many cells. Another problem is that even now there is a shortage of high-purity silicon, the raw material for making solar cells. Up until now, solar cells have been manufactured using the surplus of extremely high-purity silicon made for semi-conductor applications, but this surplus has run out. Until an alternative supply of high-purity silicon can be found, it will be difficult to increase production of solar cells. This is one reason why the contribution of solar cells is set at only 3% in Vision 2050. However, by developing an industry to manufacture even this limited amount of solar cells, we will solidify the position of solar power as an energy ace for the latter half of the 21st century.

The most common solar cells on the market today, silicon solar cells, are made by reducing the raw material silicon oxide to pure silicon, which is subsequently made into an extremely thin film just a few microns thick. The fragile film of silicon is then enclosed in a frame made of aluminum and glass. Currently, the process of reducing silicon oxide into crude silicon is done in countries where electricity is cheap. Then chemical companies and steel-making companies make high-purity silicon from the crude silicon, and electric appliance manufacturers make the solar cells. One fundamental principle for increasing efficiency that we saw in the iron and steel industry is integrated manufacturing. The same principle can be applied in solar cell manufacturing. If the steps from purification of the crude silicon to the production of solar cells were integrated into one continuous process, energy efficiency and efficiency in using raw material could be increased dramatically. In fact, a doctoral thesis from the University of Tokyo in 1999 showed that with process integration, the price of solar cells could be reduced to less than one tenth of what it was at that time.

Utilizing the Polar Regions and Outer Space

As we saw in Chapter 6, the biggest problem with wind power is its stability. But if wind farms were located at the North and South Poles, they might not suffer from this problem. Near the Poles, a wind called the *kataba* blows from the Polar Regions to the surrounding areas. Like the trade winds in the low latitudes and the westerlies in the mid-latitudes, the *kataba* wind is a global scale phenomenon created by the energy of the sun and the rotation of the earth. Unlike regional winds that blow intermittently, these global winds are steady. Although currently the *kataba* wind is not harnessed for any human purpose and so merely dissipates into heat, it has been said that this resource has the potential to supply all the energy required by human civilization today.

In developed countries there are few places to install solar cells other than on the roofs of buildings, and it is difficult, using only rooftop arrays, to generate enough power to make a large contribution to a country's supply of energy. Therefore, researchers are studying methods for setting up solar cell power plants in deserts and even on geostationary satellites. Locating power plants in remote areas raises the problem of how to transport the electricity to places where it is needed. Superconductors show promise for realizing a global network of high capacity transmission lines. Researchers are also looking at ways to transport the energy of electricity economically in the form of fuels such as hydrogen or methane.

Untapped sources of renewable energy might be easier to utilize in places where there are few people. However, producing electricity in remote areas like the Polar Regions and deserts raises other issues that must be resolved, including issues of international law and local culture. Once a plan for harnessing these sources of energy has been developed, the next step must be to form an international agreement between all of the affected nations for moving that plan to the experimental stage. Only after enough evidence has been gathered indicating that the plan will benefit all the nations affected with no harmful side effects will it be possible to proceed to full implementation.

Certainly other large-scale systems could be proposed in addition to the ones described above. And it is no easy task to decide which of these systems we should invest in. But one thing is certain: if we continue to leave such decisions to experts, bureaucrats, and entrepreneurs – who comprise only a tiny fraction of society – the result may not be what is best for society as a whole. Plans made without considering a range of perspectives are often flawed, and even a decision that could have been correct under certain conditions may not have the planned outcome without broad-based cooperation.

To establish a broad-based cooperation, we must create a forum for exchanging ideas and building consensus. Such a forum must exploit the most advanced technologies for gathering ideas and exploring them from different angles. We must evaluate not only intended consequences of a plan but also possible unintended ones, possibly by using small-scale experiments and computer simulations. Only by thoroughly examining many different ideas in such a forum can we build social consensus. In the final section of this book, let's consider the conditions needed for creating such a forum.

Designing the Komiyama House

But first I would like to tell you about another project for sustainability that is a bit smaller and, for me, quite literally closer to home. It was a project to redesign my own home. Five years ago, I decided to build a new house, and I made it my goal to see how much I could reduce the energy that I consumed in my own "daily life" activities. One of the first decisions in building my new house was to equip it with a rooftop solar cell array. At the time, the 3.6 kW solar cell system cost me 2,360,000 yen, or about 20,000 U.S. dollars. However, even in 2002, the Japanese government was offering subsidies to home owners installing solar technology. I received a rebate of 360,000 yen, so the actual cost to me was about 17,000 U.S. dollars. To this rooftop solar cell array, I added a high performance air conditioning system with a COP of 4, a heat pump for my hot water supply with a COP of about 3, and 1.4 watts per square meter per degree C of insulation. I bought new appliances with high energy efficiency. All of these investments in energy conservation cost me an additional 1,240,000 yen or about 10,000 U.S. dollars. As a result, my new home requires less than half of the energy needed to run my old home, and the solar cell array provides about two thirds of that energy. So my new 207 square meter home requires only a sixth as much electrical energy from the power grid as my old home – less than 3,000 kilowatt hours per year!

Another step I took to reduce my "carbon footprint" was to trade in my old Toyota sedan for a new Toyota Prius. The Prius, a hybrid car, cost 679,000 yen more than a comparable Toyota Corolla, a little less than 6,000 U.S. dollars. By adjusting my driving a bit with the help of the friendly dashboard interface, I reduced my gasoline consumption about three-fold. As a result, my total energy use fell from 20,800 kilowatt hours per year in 2002 to 4,000 kilowatt hours in 2008. And the total cost to me was just 3,770,000 yen, or about 33,000 U.S. dollars.

4 Rebuilding the Relationship Between Technology and Society

The Problem of Dioxins

Developing a plan based on energy and recycling to establish a civilization that can be sustained on the earth requires that we model a complex system in which multiple elements interact through many intertwined relationships. There is unlikely to be a single optimal solution. Instead, we must choose from among several solutions, each of which is almost optimal but has some particular drawbacks.

As an example, let's consider the complexity of the problem of dioxins. "Dioxin" is a generic term for a group of mainly carcinogenic chemical compounds with a complex molecular structure containing chlorine in addition to the carbon, hydrogen and oxygen found in substances such as carbohydrates. Dioxins are sometimes emitted when garbage is incinerated. But if the incineration is carried out at a high enough temperature, no dioxins will be formed.

Because dioxins contain chlorine, dioxins will be formed only if there is chlorine in the garbage at the time of combustion. One source of chlorine in garbage is polyvinyl chloride (PVC), a type of plastic with a wide range of applications. Another source is the plastic wrap used for food products, which has a similar molecular structure. Recently, there has been talk of banning the use of these plastics, but even if we stopped producing PVC, dioxins would still be created. The reason is that food refuse also contains a source of chlorine: sodium chloride or ordinary table salt. So to eliminate all sources of chlorine, we would have to exclude food refuse from garbage incinerators.

If we did ban PVC to keep chlorine out of the garbage incinerator, then another problem would emerge – we would face a shortage of caustic soda. Chlorine is produced through the electrolysis of sodium chloride. During the electrolysis of sodium chloride, chlorine is created at the anode and caustic soda is created at the cathode. Because PVC is one of the main commercial uses of chlorine, chlorine would no longer be in demand if PVC were banned. As a result, the electrolysis of sodium chloride will no longer be economically viable, and the supply of caustic soda would dwindle. Because caustic soda has many important applications, such as in making soap and in neutralizing waste water, this shortage would be a problem.

If banning PVC is problematic, we could consider replacing existing incinerators with ones able to withstand high-temperature combustion. Then we could incinerate garbage containing PVC and food refuse without releasing dioxins. But is this really the best option? If you consider the question from the point of view espoused in this book, you may ask whether we should be using fossil fuels or electricity to burn garbage containing valuable energy resources such as plastic and paper at high temperatures just to prevent the formation of dioxins. After all, you have already seen that incineration of garbage is not an efficient way to produce electricity.

So how should we solve the problem of dioxins? In the previous chapters, you have seen that it should be possible to create an energy-efficient system for circulating materials, a system that can reuse waste such as paper and plastic either by recycling or by making fuel. The key is to separate those waste materials from food refuse and other garbage. If at the collection point, plastics are separated from other garbage, this plastic waste – even if PVC is mixed in – is not so difficult to process. Technologies are already available that use heat treatment to get rid of the chlorine and then use the treated waste as a coke substitute in blast furnaces. Thus it is possible to save fossil fuel resources equivalent to the amount of garbage reused while preventing the formation of hazardous dioxins. This example suggests that by carefully evaluating the way we manage our resources, including our waste materials, we can make Vision 2050 a reality.

Currently, in Japan garbage disposal is the responsibility of the local municipalities. What if one municipality takes measures to control the emissions of dioxins by improving its incinerators? That decision, in and of itself, may not be a bad idea, but in terms energy efficiency it is far from ideal. What we discover is that a choice that may seem good on a small scale – good for one municipality – may work against constructing a large-scale system that would be even better. In today's society, problems and stakeholder interests are intertwined in such a complicated way that, with the best of intentions, decision-makers often choose suboptimal solutions. We must look at each problem from a variety of vantage points and make decisions that take into account all the related aspects – from the big picture down to the fine details. And to do this, we must set up a social infrastructure for forming consensus based on discussions that involve as many stakeholders as possible.

Structuring of Knowledge and a Place for Debate

To make Vision 2050 a reality, it is essential to develop and introduce new technologies. It is no overstatement to say that only when there is a good relationship between society and technology will the sustainability of the earth become possible. But recently some people have come to see technologies as the contents of a

Pandora's Box opened by science and released upon humanity, causing misery and destruction. When we remember that science gave birth to the atomic bomb, has contributed to the destruction of ecosystems, and has given us the power to manipulate human life, it is understandable why some people may hold this perception.

Therefore, to pave the road to Vision 2050, scientists and engineers must take the initiative in starting a dialogue with society about technology. In this dialogue, we must guarantee a high level of transparency about scientific findings and must fully disclose to the public the known results of research and the likely consequences of development of different technologies for a sustainable earth.

I would like to tell you about an incident of public disclosure about technology, an incident I was involved in several years ago. In the early 1990's, the Japanese government funded a project to develop a computer program for calculating the cost and energy payback times for solar cells. One of the preconditions of the project was the public disclosure of all the findings together with the methods by which the findings were made. Over the course of a year, discussions were conducted in the public venue of a research panel at the Society for Chemical Engineering of Japan. Based on those discussions, a method for obtaining the payback time of solar cells was developed, and all of the assumptions and calculation methods were made public. Anyone who had a question about the assumptions or numbers used in the calculations could change the corresponding values and recalculate the payback time. In fact, one expert, who had originally reported that the energy payback time was five years, used a computer program produced by the project to conduct a verification of his numbers and ended up agreeing that two years was almost right. This example shows how the program acted as a platform for establishing a consensus regarding the highly complex problem of calculating cost and energy payback times for solar cells.

The Internet is sure to play an important role in facilitating public disclosure of research and development. Already, it has become common for research institutes and even private companies to publish information on the web about research activities and product development. Although it takes significant effort to maintain a website with this information, experience has shown that the advantages in terms of a company's image outweigh the costs. As another example, a group of researchers at the University of Tokyo have used advanced artificial intelligence and web technologies to develop a web-based platform that lets scientists add specially formatted descriptors to their scientific publications that can be read by a computer search engine. These special computer-interpretable descriptors function like "barcodes" that help search engines and other knowledge retrieval systems on the Internet more effectively match knowledge needs with knowledge seeds. Although the platform is still at an experimental stage, the hope is that this work will lead to publishing results of scientific research in a way that is more immediately accessible to stakeholders in society. For example, a non-expert interested in learning more about state-of-the-art research on solar cells could draw on the computer interpretation capabilities to "translate" expert scientific expressions into language that person understands.

Another example of how to bridge the gaps between researchers and stakeholders can be seen in the Tokyo Greenhouse Gas Half Project (THP). This project was initiated in 1996 with the goal of drawing up a plan for reducing by one half the emission of greenhouse gases in the city of Tokyo. The core members of THP were researchers and professors from the Faculty of Engineering at the University of Tokyo, who worked in collaboration with researchers from the Massachusetts Institute of Technology and the Swiss Federate Institutes of Technology as well as other universities and research institutes in Japan and around the world. The primary objective of the project was to evaluate the potential for combinations of technologies and policies to reduce the amount of greenhouse gases generated by a range of factors, including cars, trains, homes, offices, garbage incinerators, construction sites, and manufacturing plants, focusing on the impact of interaction effects between those different technologies and policies.

This project has had one other important aim: the development and implementation of methods for effectively communicating the information necessary for a research study on the complex systems of a city the size of Tokyo. As is shown on the project web site (http://www.thp.t.u-tokyo.ac.jp/thp_en), in addition to coming up with a comprehensive plan for reducing CO_2 emissions in Tokyo, researchers in THP also considered how current methods for enabling effective information exchange between engineers and experts from different disciplines of science and technology could be extended to make possible a discussion between all kinds of people who are interested in the object of the study, including ordinary citizens, policy-makers, and experts.

In recent years, it has become evident that we need a new academic discipline – sustainability science – to address the issues above in a more structured way. An on-going example of this science at work exists in the collaborative research and education undertaken by the University of Tokyo, the Massachusetts Institute of Technology, the Swiss Federal Institute of Technology, and Chalmers University of Technology under the Alliance for Global Sustainability. In 2005, with the support of the Japanese government, the Integrated Research System for Sustainability Science (IR3S) was created at universities and research institutes throughout Japan, including the University of Tokyo. The IR3S aims to form a network in Japan for coordinating sustainability science research and education. IR3S has begun a program addressing sustainability issues led by three flagship projects: "sustainable countermeasures for global warming," "development of an Asian recycling-oriented society," and "conceptualization and development of global sustainability focusing on reform of the socioeconomic system and the role of science and technology." The University of Tokyo has also started a new graduate program in sustainability science emphasizing exercises and projects that help students master the diverse set of academic skills and practical knowledge required to become leaders in the effort to establish a sustainable global society.

As a consequence of the specialization of knowledge, even for a single field of science or technology, each expert's breadth of understanding has become extremely narrow. It is worth taking a moment to think about why this has happened. We hear about the great Renaissance Men (invariably, the people with the time and resources to become great thinkers during the Renaissance were almost all men), such as Leonardo da Vinci, Galileo Galilei and Benjamin Franklin, all masters of a wide range of disciplines both in science and the arts. Some people may say that we have become less intellectually agile in modern times. However, it is not that the modern individual's capacity for processing information has decreased in comparison to that of the Greek philosophers or the Renaissance Men. Rather, the huge increase in the amount of accumulated knowledge, which has expanded at an accelerating rate due in part to the trend in science of splitting disciplines into narrower fields since the days of Isaac Newton, is enough to overwhelm even the greatest modern geniuses. Today, even the most devoted intellectuals can hope to sample only a small fraction of the vast accumulation of human knowledge within their lifetimes. If Aristotle or Su Song were alive today, even they would find the breadth and depth of current human knowledge overwhelming.

Here is just one example. You probably remember the "Y2K problem," the fear that some erroneous computer operations would occur when the clocks built into older computers changed from December 31, 1999 to January 1, 2000. Now we may look back at the confusion and consternation during the final months of 1999 with some embarrassment, but at the time the concern was quite real. Danny Hillis, an American inventor, entrepreneur, and author, made the following thought-provoking comment regarding the real nature of the problem:

> *I have come to believe that the Y2K apocalypse is, in the truest sense of the word, a myth. It is a shared falsehood that carries within it a profound truth. ... There are no real experts, only people with partial knowledge who understand their own little pieces of the puzzle. The big picture is a mystery to us, and the big news is that nobody knows.*

This comment exemplifies the present difficulty of "increasing complexity of social problems and increasing subdivision of fields of knowledge." We must work out a method for understanding the big picture behind the problems that we face today.

So what is required in order to do this? The first step is to carry out a widespread structuring of knowledge. One problem adding to the difficulty of accessing specialized knowledge is the cryptic way in which knowledge is expressed in each specific field. As human knowledge has expanded, members of each discipline have developed their own specialized vocabularies to communicate the results of their scientific research. At the same time, scientific publications expect their readers to be familiar with an increasingly large set of specialized terms and tacit assumptions.

"Structuring knowledge" means making the specialized knowledge in specific fields clear to people outside those fields by establishing the connection of the ideas in that field with the whole of human knowledge. When scholars report knowledge that they hope to be helpful in achieving a sustainable earth or addressing some other social need, they must prune the jargon from their prose. Only then will actors in society be able to understand that knowledge and translate it into actions. The responsibility for doing this must lie with the members of each field. But even between related fields in the sciences, the same words may be used to express very different concepts, so an electrical engineer, for example, may interpret a paper written by a physicist in a completely different way from what was intended. To make their knowledge more structured and accessible, specialists need to establish clear definitions in everyday language for the terminology they have developed in their specific fields. This must be done in parallel to the process of publishing specific research findings. Most scientific disciplines have one or more representative societies, where members of the discipline gather to share ideas related to their field. These academic societies might be good places for scientists and other specialists to establish how their work is related to other fields of knowledge. To clearly describe the way knowledge in each field is connected with that of other fields, the specialists must focus on the meaning of the entire field rather than getting mired in specific details.

Computers may facilitate the difficult task of structuring knowledge. In the same way computer algorithms have been developed to translate text between languages as different as Japanese and English, it may be possible to develop computer-based techniques for translating the materials written by experts to describe their knowledge, such as papers in professional journals, from one field, such as chemical engineering, to another, such as economics. But if computers are to play the role of interpreters, the specialists must prepare descriptors of the knowledge they are sharing in ways a computer can understand most easily and most accurately. Just as we do not burden a human translator with jargon and expressions unfamiliar to the translator, these descriptors must avoid ambiguous human expressions that would baffle a computer translation program.

Another step in making accessible the "big picture" behind the large-scale and complex problems of society is finding a way to store the structured knowledge in a form people can easily tap into. Suppose that we wanted to present the latest expert knowledge on the current state of global warming, on the role played by solar cells, and on the time it takes solar cells to pay for themselves. And suppose that we wanted to present this knowledge in a way that could be accessed easily by people deciding whether to invest in a solar cell system. This knowledge should be presented in such a way that each area of related knowledge is integrated seamlessly with the overall topic: how investing in solar cell systems can help mitigate global warming. By presenting this knowledge on a web site in a way that allows feedback and dynamic interaction, the person accessing the knowledge on the web site, who may have a question about what he/she is reading, can pursue that question by interacting directly with the web site. Already several interesting web sites are providing access to expert knowledge in this way. We must continue developing the computer infrastructures and software tools that allow experts to share their specialized knowledge themselves in integrated, easily accessible formats with minimal effort.

The analysis and vision presented in this book represent an attempt to articulate an overview of the entire system of human activities within the earth's biosphere, and to use that overview as a framework for planning how by wise use of technology we human beings can assure the sustainability of the earth. Certainly this book has not included all of the specifics related to every human activity and every technology that could be included in a plan to realize a sustainable earth. To give but a single example, there is no question that experts on transportation and automobile engineering know the details regarding the design and implementation of energy efficient automobiles far better than the authors of this book. To build a sustainable future for the earth, detailed knowledge of technologies, human activities, and the workings of natural systems will certainly be necessary. But it is our belief that what we need right now is a clear and comprehensive vision of how our activities and the technologies determining how those activities are performed relate to the earth as a whole. Once we have a shared vision of the whole, we can focus on the specifics, always with an eye on how those specifics affect the entire system of human activities and what implications those specifics have on the sustainability of human life on the earth.

Postscript

There are several reasons why I decided to take on the rather immense task of writing a book proposing a macro-vision for a sustainable earth.

Of course, first and foremost is my strong belief in its necessity. It seems that governments, businesses and individuals today each take their own separate stances and act without any coordination – almost like looking at a Pointillist painting with no motifs. I believe the reason for this disjunction of key members of society is that there is no shared "big picture" among them. What we need now is not a simple compilation of details, but rather a big picture based on considering human activities and the earth's response together.

The second motivation for writing this book comes from my growing confidence that it is indeed possible to create a big picture that could be shared between researchers with different kinds of expertise or even between people without any particular expert knowledge. That big picture would presume only a small number of basic principles such as the conservation of mass and energy.

My goal in writing this book has been to communicate this kind of shared big picture, so I have avoided the use of specialized expressions. I often speak with people from the humanities about environmental topics, and the major obstacle in communicating with them is differences of expression. If the meaning of just one expression, such as a word, an equation or a specialized concept, is not understood, it is impossible to understand the overall idea being communicated.

For example, entropy is a fundamental concept of thermodynamics and energy, but in fact there are few people, even among experts in technology, who really understand its meaning. However, it is possible to discuss the principles of energy efficiency without going into the details of entropy. I have attempted to take this approach in writing this book.

The 21st century is the era in which human will determine the future of the earth and society. Therefore, we must not let society develop a misunderstanding of science and technology, whose power to influence the world has grown enormously. However, even today countless misunderstandings still go unchallenged.

One obstacle to understanding new findings and developments of science and technology might be a conscious effort on the part of the experts to try to make their work look as difficult as possible. When I gave a lecture on my research at a reunion some years ago, a friend of mine said to me "I was surprised to hear how simple the things that university professors do are." Of course I was not particularly thrilled to hear my work described as "simple," but what point is there in giving a talk that no one understands?

Even without this conscious effort of some researchers to make things sound difficult, we are faced today with a fundamental difficulty emerging from increasing complexity of problems we must address and a simultaneous increase in subdivision of fields of expertise. For this reason, the responsibility of experts to explain their fields of expertise in simple terms is all the more important. We can leave the details to the experts of that field. However, it is necessary to transcend the individual fields of expertise to achieve an understanding of the fundamental overall structure.

The third motivating factor for my decision to write this book is the support of an uncountable number friends and acquaintances. Today, unlike the times of the ancient Greek philosophers, no one person can grasp the sum of all things known to humanity. The amount of knowledge that has been gained by humanity has exceeded the information processing ability of humans. If I am to speak of a macro-vision, there is no way to avoid having to touch on fields of expertise outside of my own. However, even if I personally am unable to understand the details in each of those fields, if I can understand the words of the experts from those fields, it should be possible for me to create a useful big picture. Knowing that I can obtain accurate information through a network that reaches beyond fields of expertise and national boundaries has been a critical factor in this difficult undertaking.

These are the reasons that have motivated me to put forth a macro-vision for a sustainable earth. That there are some errors in the details behind this big picture is unavoidable. Certainly we must listen to and learn from the criticisms of experts from a wide range of fields. In fact, by exposing Vision 2050 to criticisms, modifying it, and filling in the details further, it should be possible to construct an even better big picture. It is my sincere hope that this process will occur.

Although I have presented Vision 2050 as a big picture, this vision is focused on materials and energy. I have intentionally left out the problem of lifestyle, and I have scarcely touched on topics related to social institutions. And even after restricting my study to materials and energy, I have had to make some important omissions. For example, this book has not treated issues related to the sustainability of agriculture. As we saw in figure 1-3, from the middle of the 20th century, agricultural production increased continuously with no hint of slowing. On the other hand, crises caused by soil runoff and degradation have been reported, and recently organic agriculture is being promoted as a possible countermeasure for these problems. However, it seems that there is little if any discussion regarding the ability of organic agriculture to supply a sufficient amount of food to meet the needs of the global population. Similarly, I have not discussed the issues related to water or the preservation of species in this book.

Just having this kind of shared overall vision is of course not enough to enable the formation of consensus for what specific actions to take in order to achieve a sustainable society. However, together with tools for aiding in the structuring and sharing of knowledge, it could form the starting point for true society-wide consensus building by providing a common understanding leading to the establishment of a global forum or meeting house for dialogue and collaboration manifested on the Internet. By creating this kind of meeting house, it is my heartfelt wish that we will be able to rebuild a good relationship between technology and society, and in doing so, chart a course to a sustainable earth.

There is a movie entitled "On the beach." At the last scene, when humanity is on the verge of extinction after a nuclear war, in the midst of dried leaves blowing about in the gusty wind, on the leaning gravestone in a church were written the words "there is still time, brother." The same is true for global sustainability. It is not too late if we take the first step now.

Hiroshi Komiyama

About the Authors

Hiroshi Komiyama became the 28th president of the University of Tokyo in April 2005. Dr. Komiyama specializes in chemical engineering, global environmental engineering and structuring of knowledge. He received his Bachelor's, Master's, and Doctoral degrees in chemical engineering from the University of Tokyo. From 1973 to 1974, he was a post doctoral fellow at the University of California at Davis. He became a professor of Department of Chemical System Engineering at the University of Tokyo in 1988. After serving as Dean of the School of Engineering from 2000 to 2002, he was appointed as vice-president of the University of Tokyo in 2003 and executive vice-president in 2004.

As President, Dr. Komiyama announced the UT Action Plan in 2005, summarizing key initiatives for realizing "the University of Tokyo that aims at the pinnacle of the global knowledge at the forefront of the time." In addition, he has initiated new projects to reform the University of Tokyo through the endeavors to achieve "structuring of knowledge" and "autonomous and decentralized yet cooperative" sharing of knowledge.

Steven Kraines is an associate professor in the Science Integration Programme of the Division of Project Coordination at the University of Tokyo. Dr. Kraines specializes in web-based technologies for expert knowledge sharing, agent-based collaboration systems, and model integration. He received his Bachelor's degree from Oberlin College and his Master's and Doctoral degrees in chemical engineering from the University of Tokyo.

Other Books by Hiroshi Komiyama

(In Japanese)

Sokudoron (Rate processes)	Asakura	1990
Chikyu ondanka mondai handobukku (Handbook of global warming)	IPC	1990
CVD handbukku (CVD handbook) *translated into Korean (1993)	Asakura	1991
Biryushi handobukku (Microparticle handbook)	Asakura	1991
Chikyu kankyo no tameno chikyu kogaku nyumon (Primer of global engineering for global environment)	Ohmsha	1992

Hiroshi Komiyama and Steven Kraines
Vision 2050: Roadmap for a Sustainable Earth.
© Springer 2008

Chikyu kankyo no tameno kagakugijutsu nyumon (Primer of chemical technologies for global environment)	Ohmsha	1992
Chikyu ondanka mondai ni kotaeru (Answering the global warming problem)	University of Tokyo Press	1995
Hannou kogaku (Reaction engineering)	Baifukan	1995
Nyumon netsurikigaku (Introduction to thermodynamics)	Baifukan	1996
Chikyu jizoku no gijutsu (Technologies for global sustainability) *translated into Chinese (2006)	Iwanami	1999
Taiyokou hatsuden kogaku (Photovoltaic engineering)	Nikkei BP	2002
Biomass Nippon (Biomass Japan) *translated into Chinese (2005)	Nikkan Kogyo Shinbun	2003
Ugoke! Nippon (Move it Japan!)	Nikkei BP	2003
Chishiki no kozoka (Structuring knowledge) *translated into Chinese (2005) and Korean (2008)	Open Knowledge	2004
Todai no koto oshiemasu (I will tell you about the University of Tokyo)	President	2007
Kadai senshinkoku Nihon (Japan as a forerunner for addressing emerging problems in the world)	Chuokoron-Shinsya	2007
Sustainability he no chosen (Taking on the challenge of sustainability)	Iwanami	2007
Chishiki no kozoka koen (Lectures on structuring knowledge)	Open Knowledge	2007
(In English)		
Equilibrium and reaction rate. In: Tominaga H, Tamaki M (eds) Chemical reaction and reactor design	Wiley	1997
New techniques to produce functional materials: chemical vapor deposition. In: Garside J, Furusaki S (eds) The expanding world of chemical engineering	Gordon and Breach Science	1994
Global sustainability and the role of Asia. In: Sasaki T (ed) Nature and human communities	Springer	2004